THE SWAN

No 736

Joanna Greenlaw

THE SWANSEA COPPER BARQUES AND CAPE HORNERS

by

JOANNA GREENLAW

Published by the Author

Published by
Joanna Greenlaw, 59 King Edward Road
Swansea, SA1 4LN
Telephone: (01792) 425494

ISBN 0 9531549 1 2

*Printed in Wales by
Dinefwr Press
Rawlings Road, Llandybie
Carmarthenshire, SA18 3YD*

TO JENNY,

In remembrance of the evening of 3rd October 1997, when halfway through a bottle of red Californian wine in John's Grille in San Francisco she suggested that I write this book.

'As for myself, the wonderful sea charmed me from the first.'

(Joshua Slocum, *Sailing Alone Around the World*)

Contents

JOSEPH CONRAD, REMINISCING ABOUT ONE OF HIS OLD CAPTAINS:

It appeared he had 'served his time' in the famous copper ore trade of the old days between Swansea and the Chilean coast, coal out and ore in, deep-loaded both ways, as if in wanton defiance of the great Cape Horn seas – a work, this, for staunch ships, and a great school for West Country seamen. A whole fleet of copper-bottomed barques, as strong in rib and planking, as well-found in gear, as ever was sent upon the seas, manned by hardy crews and commanded by young masters, was engaged in that now long-defunct trade.

D'ye mind the time when we squared away and ran her
East by South,
When she trampled down the big Horn seas with a
roaring bone in her mouth,
When the best hands twirled her bucking wheel an'
dared not look behind
At the growling greyback in her wake? Dy'e mind,
old pal, d'ye mind?

F. W .WALLACE

Quoted by Basil Lubbock in his
Last of the Windjammers, Volume I

Introduction

IN HIS INTRODUCTION to the facsimile reprint of W. H. Jones's *History of the Port of Swansea*, Professor Sir Glanmor Williams wisely remarked that all students of history stand on their predecessors' shoulders. This has been constantly in my mind, for it is impossible for anyone to write about the Swansea Copper Barques and Cape Horners without being aware of the writings of local historians like W. H. Jones, and marine historians like Percy Lubbock, of illustrious memory, Basil Greenhill, Aled Eames, and Michael Stammers, to name but a few.

I am only too well aware that each of the chapters of this book could fill a whole volume, and it is against this context that I have endeavoured to present an overall picture of Swansea Port, its neighbouring towns of Llanelly and Neath, and its merchants, industrialists, copper barques and Cape Horners.

Wherever possible I have gone back to original sources; it has been necessary to bring together many threads: the Victorian owners and entrepreneurs who built and sent the vessels to sea; the building of the ships; the men who sailed in them, and the old seaport towns of Swansea, Llanelly and Neath.

The story shifts to Cape Horn, with all its challenges, horrors, and tales of almost unbelievable courage, seamanship, and human tenacity in the face of the worst weather nature could do at the bottom of the world.

Then to the west coast of South America, with its copper, nitrate, and guano ports straggling along the long seaboards of Chile and Peru, alive with drinking dens, brothels, seamen's boarding houses, crimps, remittance-men, beach-combers, and adventurers of every kind.

And then San Francisco itself, queen of the north Pacific, created by a heady mixture of sailing ships, gold, and as one end of the Oregon trail.

They are all gone now; the ships, the men, and the risks they took; all we have to remind us are museums and those old sepia photographs of ships and crews taken in ports all over the world. They stare out at us, tough, confident, defiant, humorous and, in a slightly sad way, socially vulnerable. They are a reminder of a vanished world which though it offered hardship had something much harder to find today: challenge and adventure for the individual, and a chance of real wealth for the few lucky ones who became owners of ships or large enterprises ashore.

For some of us today, among our washing machines and computers, the name Cape Horn sets our blood racing and creates in us a surge of that human energy which drove ships around the world under the power of sail alone. And what better way to describe them – the crews rather than the officers – than by Joseph Conrad in his immortal 'Nigger of the *Narcissus*':

> "But at the corner I stopped to take my last look at the crew of the *Narcissus*. They were swaying irresolute and noisy on the broad flag-stones before the Mint. They were bound for the Black Horse, where men in fur caps, with brutal faces and in shirt sleeves, dispensed out of varnished barrels the illusions of strength, mirth, happiness; the illusion of splendour and poetry of life, to the paid-off crews of southern-going ships . . . The crew of the *Narcissus* drifted out of sight. I never saw them again. The sea took some, the steamers took others, the graveyards of the earth will account for the rest.
>
> Good-bye, brothers! You were a good crowd. As good a crowd as ever fisted with wild cries the beating canvas of a heavy fore-sail; or tossing aloft, invisible in the night, gave back yell for yell to a westerly gale."

In asking myself why, in my seventies, I embarked on the daunting amount of research, travel, expense, and sheer slog that a work of this nature demands, my answer has to be that of Sir Richard Burton, Victorian explorer, soldier, poet, translator, geologist, swordsman, author and raconteur: 'Damned Fool! the Devil drives.'

If there are any faults in this book the responsibility is entirely mine.

1.

Cape Horn

THE PATAGONIAN ARCHIPELAGOS – the numerous islands fringing the western coast of the mainland, between Magellan's Strait and the Penas Gulf form the Patagonian archipelagos: these, like the southern and western parts of the Archipelago de Tierra del Fuego, are about as inhospitable as anywhere on the globe. The land is mountainous, presenting an alternation of impenetrable forest, bare rock, and deep bogs, and is cut up by deep channels into peninsulas and islands, as yet very imperfectly known. Drenching rains, varied by snow and sleet, prevail throughout the year, whilst furious westerly gales succeed each other with rapidity. The scenery is magnificently stern, but is seldom seen to advantage, the clouds and mists usually screening the higher peaks and snow fields: glaciers, however, extend in many places nearly, or quite, to the level of the sea. In such a climate life is scarce.

The Admiralty South America Pilot – Volume II

They that go down to the sea in ships and occupy their business in
great waters,
These men see the works of the Lord, and his wonders of the deep.

Psalm 107

Cape Horn is the most desolate and feared cape in the world, and its awesome reputation derives from its peculiar geographical position. Situated at the bottom of the world, the prevailing winds in the forties and fifties south blow constantly from west to east, and in winter the storms rarely cease. A succession of depressions roll eastward, generating waves which can build up as high as 100 feet or more, uninterrupted by any land mass such as occurs in the northern hemisphere. They represent perhaps the greatest continuous natural force to be encountered anywhere on earth.

The Andes, one of the longest and greatest mountain ranges in the world, block off the westerly winds along a front of over a thousand miles, and all this force is squeezed through Drake's Straits between Cape Horn and the South Shetland Islands, five hundred miles to the south.

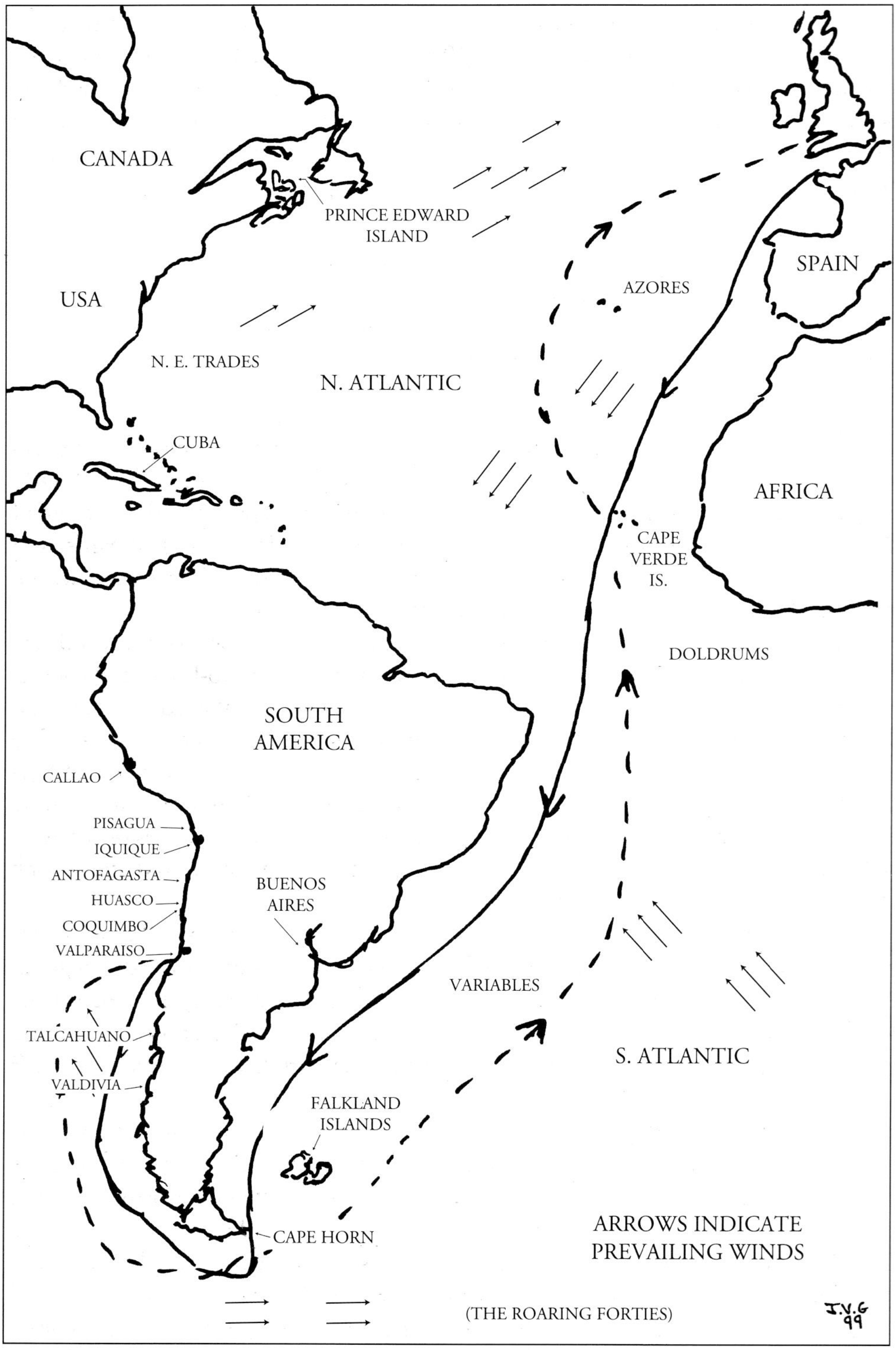

Sketch map showing the Cape Horner's route.

Conditions were not quite so bad if the ship was sailing from west to east; the 'Roaring Forties' (the latitudes between 40 and 30 south) acquired their name from the countless sailing ship masters who took advantage of these winds to blow them, sometimes heeling with one deck side awash for weeks on end, from the Cape of Good Hope to Australia and New Zealand. As the old blue-water seaman's jingle went:

> Down the Channel, Cross the Line,
> Well south of Table Bay,
> Round the Horn and home again.
> That was the sailor's way.

But the passage from west to east also presented horrific dangers; instead of weeks of tacking against westerly winds there were huge following seas to contend with, which could rake a ship from end to end, sometimes sweeping the decks clear of boats, galley and housings. They used to call these gigantic waves greybeards, and the helmsmen, (because sometimes two men were needed) lashed to the wheel, never dared to look back at the seas which were following them.

Basil Lubbock quotes one experienced Cape Horner: 'The force of the wind was so great that the ocean smoked, and one could not see the jib-boom for spume, which flew through the air like steam'. It is probably impossible for anyone who has not experienced the destructive violence of a Cape Horn storm to visualise them. The furious westerly squalls create huge seas which look like a range of hills following the ship. Now and again, one larger and higher than all the rest towers over the ship's stern; its head turns a translucent green, and it crashes down with solid force upon the ship. Such breakers can twist iron and steel, and smash oak and teak as if they were matchwood.

But it was the outward passage that was feared most of all, with the fierce rip-tides of the Le Maire Strait, fringed by the iron-bound rocks of Cape San Diego on one side and Staten Island on the other. Men could be critically injured on the flooded maindeck of their own vessel, let alone being washed overboard or falling from a yard foot-rope after hours spent aloft fighting with heavy canvas frozen stiff with ice.

The very names of the capes and inlets give an idea of the impression they made on the minds of those who named them: Cape Desolation, the Furies, False Cape Horn, Dislocation Harbour, Cape Deceit, Clear Bottom Bay, and Christmas Sound. As ships approached Cape Horn the crew would prepare the ship for the battle ahead, knowing they would have to face the worst storms a seafarer could encounter. The standing and running rigging would all have to be checked and overhauled where necessary, life-lines stretched across the main deck, and hatches securely battened down. Bags of oil were carried, which could be released with the effect of

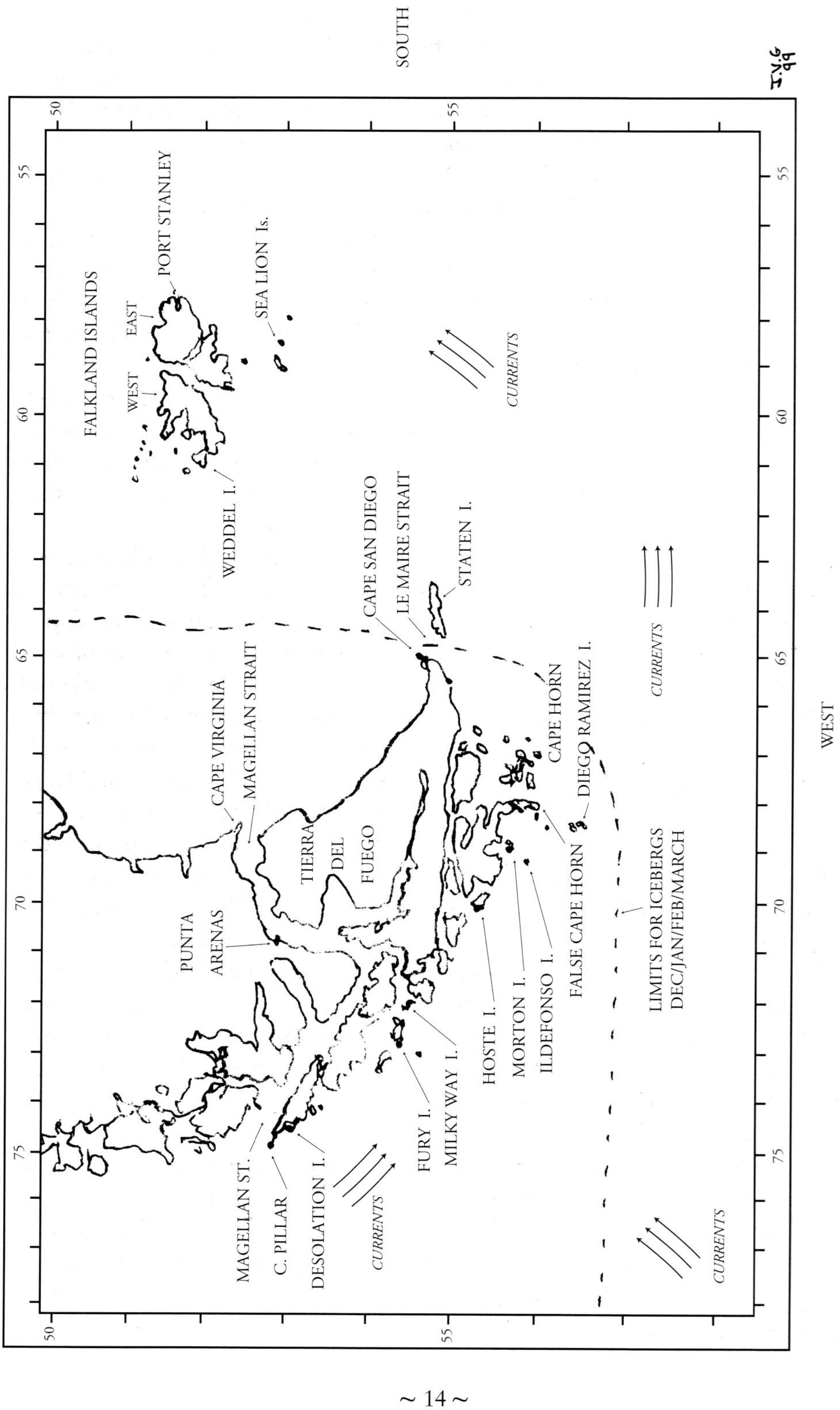

Sketch map of Tierra Del Fuego and Cape Horn.

calming down a following sea. Sometimes a hawser would be trailed astern in a long bight, which acted as a sea-anchor to bring the ship's head into the wind.

When rounding Cape Horn from east to west, sailing ships had to constantly tack and beat their way well to the westward before turning to head north. There was always a strong current running to the east so that shipmasters always flogged their way to windward, often with the sails and rigging frozen with ice, until they had cleared enough to westward to go round on the opposite tack.

The average passage was probably two or three weeks, but it sometimes extended into months, with food and water running short, and no hot meals or drinks because the galley fire was doused by saltwater. Sometimes vessels gave up in despair or were so badly battered that they were lucky if they could make it as far as the Falkland Islands for shelter and repairs.

As the 19th century wore on, it was not uncommon for twenty or thirty vessels to be within sight of one another off the desolate shores and snow-covered mountains of the Tierra del Fuego (Land of Fire – a wry name). Many got no further, and today the hulk of the wooden three-masted full-rigged ship *Jhelum* lies in Port Stanley harbour, a mute testimony to the force of the southern seas.

The writer has in her possession an historical plan of the Falkland Islands and their surrounding waters drawn by Mr John Smith, MBE, FRGS, now curator of the Falklands Museum, Port Stanley, which she acquired while serving on a ship in the Falkland Islands in 1982. On it Mr. Smith has marked the position of over eighty sailing vessels which fell victim to the Cape Horn seas. The exact number of vessels lost while trying to round Cape Horn is not known. In 1905, fifty-five vessels were posted missing in Britain. They sailed and were never heard of again.[1]

It was not uncommon for vessels, after weeks of fruitless efforts to round Cape Horn, to turn and sail right around the world and across the Pacific from the east. Calms were always a nuisance to sailing ships, but off Cape Horn they presented a special menace of their own. Occasionally, the strong westerly wind would cease altogether, leaving behind it sixty feet high waves with sides like bottle green glass and crests that, almost in slow-motion, toppled over before breaking.

With no wind to provide steerage way, a ship could dismast herself or roll right over, or end up in a trough between two waves, with one or both threatening to crash down on the deck with a force of hundreds of tons. And even if the sea was relatively calm, the vessel was exposed to worse danger – that of being helplessly carried by the tides and currents on to the rock-bound Cape Horn coast. But there were still other hazards.

Every year immense sections of the Antarctic ice barrier break off and drift in a north-easterly direction into the South Atlantic. Huge floating fields of ice are thus formed, sometimes in the shape of a horse-shoe 60 miles long in one direction and 40 in the other. As these break up into individual icebergs, the menace they present

1. Alan Villiers, *The War with Cape Horn.*

to a sailing vessel is not difficult to imagine. In May 1892 the ship *Kinfauns* sailed through over 400 miles of ice and counted no less than 100 bergs in one morning. The colours of some of these Antarctic icebergs are often incredibly beautiful, sometimes reflecting a dazzling white in the brilliant sunshine, with shades of blue, green and orange near to the water's surface.

Then there were the Cape Horn fogs, and nothing is worse at sea than a thick fog. Sounds lose their direction, and on everybody's mind was the possibility of being carried by the currents on the rocks or, just as bad, icebergs. And all hands knew that as soon as the calm lifted it would be followed by another storm.

Captain William Nelson, commanding the *Acamas* round the Horn in 1905 – '. . . a voyage that included sheltering for several days in the lee of Staten Island off Tierra del Fuego from an extra savage gale – saved his men from frost bite by having a tub of straw lashed beside the wheel for the helmsman to stand in.'

Later, during a hurricane, he records 'the two best helmsmen on board were kept at the wheel, getting a small tot of rum periodically to warm them . . . A high dangerous sea swept the main deck as the ship staggered along about twelve knots under lower topsails . . .'[2]

No exact account is available of how many vessels were lost round the Horn; the first on record was probably Drake's pinnace. In the days before radio the last heard of a missing vessel would be a sighting by another ship, or a tale told by survivors who were either picked up by another vessel or managed to struggle ashore. Survivors in a ship's boat, probably with little food or water, had a grim task ahead of them; the Cape Horn currents would tend to carry them eastward into parts of the Southern Ocean well away from the usual sailing ship routes. South Georgia and its whalers was probably their only hope.

Another factor was the competition between masters for quick passages; voyage times were often the subject of large wagers, and this encouraged masters to drive their ships and crews to limits of endurance. German masters had a bad reputation with crews for this, and seamen tended to avoid sailing with them if they could find another ship.

It was no wonder that young apprentices returning from their first experience of doubling Cape Horn had rousing tales to tell their families and girl-friends, and how proudly they would produce their tobacco pouches made of albatross skin, or a pipe whose stem had been cut from a wingbone.

In old age the Cape Horners invariably looked back on their years at sea with fond memories. Those who survived drowning and injury tended to live very long lives – not surprisingly in view of the physique they must have had in order to endure the hardships encountered on southern going vessels.

In 1957 Captain C. Haag of Darmstadt, writing in *Sea Breezes*, looked back on his youth on a Cape Horner in the 1880s as follows:

2. Hugh Falkus, *Master of Cape Horn*.

Cape Horn.
(Photograph by Bill Kooiman, San Francisco)

The steam auxiliary barque 'Zeta'. School of Samuel Walters, Liverpool.
(Private Collection)

Taking on the pilot, off Mumbles Head, by James Harris.
(Private Collection)

A view of Cameron's Wharf, Swansea, circa 1847. Attributed to William Butler (1823-1870).
(Courtesy Swansea Museum)

> "Those happy reckless days of youth I now recall in my old age. They were also heart breaking-days, when a man on board ship was judged alone on his face value and ability to do the work. They were the days of wooden ships and iron men, the days when a man who was hanging on with you to a lee-forebrace, waist deep in ice-cold water, 40 degrees down South, bound around the Horn, was considered your blood brother, irrespective of his nativity or creed.
>
> And never shall I forget the days when the ports of Hamburg, Liverpool, San Francisco and Shanghai were full of sailing vessels; when in Sydney and Melbourne dozens of them were waiting for their wool cargoes, and old Charley Moss in Sydney and the Rev. James Fell in 'Frisco gave the boys a good time in the sailor's missions. When I counted 92 ships in Iquique harbour, each one of them gave us three rousing sheers on the night we hoisted the crossed red and white lights, with a shanty, after taking over the last bag of saltpetre.
>
> And I shall never forget the haunting tune of the chorus, when the outward bounder broke out his anchor on sailing day:
>
> Rolling home, rolling home,
> Rolling home, across the sea
> Rolling home, to thee, dear homeland,
> Rolling home, my land, to thee!"

The seamen who manned these ships must have been probably the toughest and most skilled in their profession who have ever lived. Their conditions and pay were appalling, but there was a camaraderie among them that money could not buy, and to be a Cape Horner in the days of sail was the highest accolade a seaman could have. And Swansea can share with New York and San Francisco the honour of having more of them than anywhere else on earth.

Perhaps it should be left to Joseph Conrad, that great seaman and even greater writer, to sum up his memory of a Cape Horn storm:

> "There is another gale in my memory, a thing of endless, deep, humming roar, moonlight, and a spoken sentence. It was off that other Cape which is always deprived of its title as the Cape of Good Hope, is robbed of its name. It was off the Horn. For a true expression of dishevelled wildness there is nothing like a gale in the bright moonlight of a high latitude.
>
> The ship, brought-to and bowing to enormous flashing seas glistened wet from deck to trucks, her one set sail stood out a coal-black shape upon the gloomy blueness of the air. I was a youngster then, and suffering from weariness, cold, and imperfect oilskins which let

water in every seam. I craved human companionship, and, coming off the poop, took my place by the side of the boatswain (a man I did not like) in a comparatively dry spot where at worst we had water only up to our knees. Above our heads the explosive booming gusts of wind passed continuously, justifying the sailor's saying: 'It blows great guns'. And just from that need of human companionship being very close to the man, I said, or rather shouted:

> 'Blows very hard, boatswain.'

His answer was:

> 'Aye, and if it blows only a little harder things will begin to go. I don't mind as long as everything holds, but when things begin to go it's bad.'

The note of dread in the shouting voice, the practical truth of these words, heard years ago from a man I did not like, have stamped its peculiar character on that gale."

THE LOSS OF THE *SAN RAFAEL*

As an example of the horrors of being wrecked on Cape Horn, the fate of the survivors of the *San Rafael* of Liverpool must surely be one of the most tragic. The *San Rafael* was bound from Liverpool to Valparaiso with coal when on the 4th January 1876 the cargo caught fire and the vessel had to be abandoned on latitude 53S longitude 76W. The story reached Swansea when it appeared in the *Cambrian* newspaper on 10th November 1876. The crew abandoned ship in two lifeboats; one of which with eleven men on board was rescued by the *Yorkshire* bound from Melbourne to Gravesend. The other life-boat, containing Captain McAdam and eight men, managed to reach land somewhere between Waterman and Henderson Islands, Tierra de Fuego. It was not until the 22nd April that a party of Indians brought news to the Rev. T. Bridge, Superintendent of the South American Missionary Society in Tierra de Fuego, that a party of Indians had heard that strangers had landed on an island, and most were dead. The Indians had found seven men dead and two still alive. They lit a fire and gave them water and a shag (a seabird) and left them promising to return the next day, the survivors being too weak to move and in any case they could not get their canoes ashore due to the kelp and rough seas. When they returned the next day both survivors were dead. On 18th May a Captain Willis of the *Allen Gardiner* accompanied by his two seamen visited the scene of the tragedy and his account in the *Cambrian* speaks for itself:

"We scrambled up the steep rocky coast, and some 50 yards above the sea, near the middle of the islet, or rather peninsula, found nine bodies much decomposed, but all abundantly clothed. One of these was the captain's wife. Seven bodies were lying in a long hollow between rocks about 4 ft. wide and 5 ft. deep and about 40 ft. long. Two bodies were away from the rest, but close to each other. At the upper end of this hollow the poor fellows had lived. Across the rocks they had spread canvas, perhaps their boat's sail, which had attracted the Indian's sharp sight from the other side of the sound. Here they had made their fires and cooked their food, tins of meat and such mussels as they could gather on the steep wave washed coast of their circumscribed prison, a peninsula about 150 feet high, 100 yards wide from west to east, and 300 yards long from north to south. It appeared from the sea to be joined to the main, but on ascending we were horrified to find an abrupt chasm completely cutting off retreat from this dreadful prison. For clothing and other things we ascertained the names of eight of the dead as follows: Mr. and Mrs. McAdam, the latter by birth we supposed to have been named M. Proctor, finding that name on an article of dress, W. H. Swift, W. Daly, G. G. Stewart, T. Umlandt, and James Reid, address of latter Lockstill or Lockhilt. Captain Willis fortunately came across a private note written in pencil on four loose leaves of a pocket book by Captain McAdam, addressed to his son. It was written on the forty-first day of their stay on the desolate island, when the writer was nearly blind and scarcely able to see the paper he wrote upon. The note was found under a little bush and stated that they were in a low state, but gave no other general information. We buried up the dead as we found them, as it was impossible to remove them. Since then we have recovered from the natives a chronometer and case, which was stowed away in a rock on an islet in New Year's Sound, a barometer and sextant case, and part of 'South American Pilot', also hid away in a like manner. A night binocular was also freely given up, and among the natives we saw various trinkets they had obtained from watches, etc. From the quantity of things we saw on the peninsula and heard of as having previously been removed by the natives, we infer that the poor fellows had leisure in removing from their vessel. There were some five or six cans, about 24 empty meat tins, odds and ends of other books. Whilst we were hunting about, the two sailors in the boat pulled up into the bay near by, and found parts of two name boards of the *San Rafael* which he suppose to have been the thwarts of the boat, the wreckage of which was lying around. The poor men could not have saved their boat on the place they landed, as they could neither have anchored it off nor hauled it ashore."

(*Cambrian*, Nov. 10, 1876)

ATTACK ON THE CREW OF THE *PRINCESS LOUISA*

There is plenty of evidence that survivors from shipwrecks landing on Patagonia were rather more likely to be robbed or murdered by the natives than helped:

> In March 1871 the steamer *Princess Louisa* sailed from Scotland for a South American port and the captain, who had heard of the murder of the officers and crew of the British brig *Propontis* sometime earlier, decided to call in at Punta Arenas to obtain arms, with the object of revenging the murders. While off Port Gallant, a cove in the Straits of Magellan, two canoes and a European-built boat filled with natives pulled off from the shore towards them. The captain had no doubt that they were the same natives who had carried out the murders, and he ran down one of the canoes, drowning all on board. The boat came alongside the *Princess Louisa* and its occupants attempted to board; but all were killed.

The *Princess Louisa* sailed on, but that night ran ashore, and was wrecked itself. The crew landed easily, taking stores from the ship and erecting a tent. Just before dawn, while this was going on, they were attacked on the beach by natives, and what occurred thereafter is best described by the only survivor, a young American called Joseph Lorritz:

> "... were suddenly set upon by a horde of savages, armed with spears, bows and arrows, rough hatchets and stones. A desperate fight ensued. The sailors defended themselves without avail; two of them were killed on the spot, and the rest were overpowered. Lorritz ran for his life, but was caught, tied and laid on the ground. His companions were served in the same way. For three weeks they all remained in captivity, allowed to wander about with their captors during the day, tied at night, and fed on what Lorritz called 'blubber'. At the end of these three weeks Lorritz one night contrived to shuffle off his bonds. He crept down to the beach, got into one of the boats lying there, and put off into the channel, where next morning he was fortunate enough to fall in with the schooner *Hutchinson*, on board of which he has come to this city. The fate of his companions, who he says were all Scotchmen, is unknown to him, and he is unable to tell more than the Christian names of the captain and crew."

(*Cambrian*, 30/6/1871)

LOSS OF THE *MANCHESTER*

On 7th April the American ship *Manchester* of Nantucket left New York for Valparaiso. On the 28th August the ship struck a sunken rock to the westward of Cape Horn at about 55 south. Filling with water, and with the main mast cut away, the *Manchester* drifted until she grounded on a small island, and soon afterwards broke into two parts. The captain and his wife and son clung together, but his wife became separated and only he and his son reached shore. Four days afterwards the second mate, David Reese Evans of South Wales and one seaman arrived on the island on a small raft they had constructed at another island about 2½ miles away. On 19th November they were attacked by Indians and the captain bludgeoned to death. The three survivors escaped in their ship's boat, taking refuge further along the coast. After six weeks, their provisions exhausted and subsisting on shell fish, they were rescued by a passing steamer, and eventually taken on board a warship, the *Metero*, and conveyed to the USA.

2.

The Copper Trade

COPPER IS THE EIGHTH most abundant metal on earth, and like gold and silver occurs in its natural state. Early man found that it could be beaten into sheets and fashioned into pots, tools or weapons. When polished, it provided a serviceable mirror, and was excellent for making jewellery.

Smelting, the basic process of heating the ore so that the copper melts and runs out, introduced the age of metals. Even primitive bellows could produce the high temperatures required. A furnace was used, and during the 18th century the heating and melting process had to be repeated several times to remove the impurities. By the middle of the 19th century furnace technique had improved to reduce the number of stages to two and eventually one. It was soon discovered that when mixed with tin or zinc stronger alloys appeared – namely bronze and brass. Basically, copper is indestructible; and it has the advantage of being easy to melt down, and re-use.

By the 1650s, copper had come into popular industrial use, and was particularly valuable in boat and shipbuilding where nails, rivets and bolts were needed in great quantities. By the mid 1700s the navy was experimenting with copper sheathing as a means of combating the damage caused to ships' bottoms, especially in tropical waters, by the teredo worm and barnacles, and by the end of the century its use for this purpose was universal.

One problem was the galvanic reaction produced by the interaction of iron, copper and salt water; if iron fastenings were used, the copper plates were liable to fall off. The early copper bolts were not strong enough to be driven through the timbers. Eventually, by the 1800s, a rolling process became available which produced copper tough enough to overcome such problems.

In the UK, deposits of gold, lead, tin and copper were found on the western side of the island, roughly on a north-south line between three and five degrees west of Greenwich embracing Scotland, Lake District, Wales, Devon and Cornwall.

During the 18th and 19th centuries, Cornwall was the leading producer of copper ore in the world, reaching a peak in 1860 of over 180,000 tons. Newcomen patented his steam engine in 1705, and by the end of the century more than seventy engines were working in the West Country. But by 1880 British copper production had dropped to 40,000 tons and by 1913 it was less than 3,000 tons.

In the meantime, the 'Great Lode' discovered in North Wales had provided Cornwall with severe competition. There were other deposits, in Scotland, the Lake District, Cheshire, Staffordshire and Shropshire, but none really achieved much more than a local importance.

Its geographical proximity to Cornwall and ample local supplies of coal needed for smelting led almost inevitably to the establishment of copper smelting in Swansea and its surrounding districts in the 17th century. By the middle of the 19th century 90% of the world's copper was being smelted in the Swansea Valley and the neighbouring areas of Llanelly and Neath. Vessels could sail up the river estuaries and land their cargo of copper ore directly into the smelting works, situated on the banks.

Lane's Llangyfelach copper works, not far from the present Landore railway viaduct, was one of the earliest built in 1717, and in 1724 Robert Morris came from Shropshire to manage the works, and subsequently became the founder of one of Swansea's leading industrial families, the Morris's (see page 55).

At the end of the 18th century there were seven copper-smelting works operating in the Swansea Valley, and one at Penclawdd on the tip of the Gower peninsula. John Vivian, 1750-1826, founder of the equally important Swansea family, the Vivians, first entered the smelting industry at Penclawdd in 1800, in partnership with the Cheadle Brass Wire Company. In 1810 he built his own works at Hafod, and the Vivian family in the persons of John Henry Vivian (1785-1855) and Henry Hussey Vivian (1821-1894) came to dominate the copper industry, not just of Swansea, but the world.

Another important firm of copper-smelters were the Williams family of Cornwall, and their Morfa works, virtually next door to the Hafod works, and built in 1834, eventually became the largest in Swansea. There was considerable rivalry between the Hafod and Morfa works.

The end product took the form of ingots of copper, which provided the raw material for a large cross-section of Victorian industry. Piping and components for steam engines were high on the list, as was copper-sheathing for the bottoms of vessels.

In addition to engineering works, the great manufacturing centres like Birmingham, London, Bristol and Liverpool consumed huge quantities of copper in the process of making goods for domestic use on a massive scale, both for home consumption and increasingly for export.

As supplies from Cornwall and the rest of the UK began to dry up or become uneconomical, foreign imports took their place; first from the near continent, like France and Spain, then Cuba, and Chile, followed by South Africa, Canada and Australia.

Working conditions were horrific, but it is important to record that the leading Swansea employers like the Morrises, Baths, Vivians and Grenfells were among the most enlightened employers of their period. They have to be judged by the standards

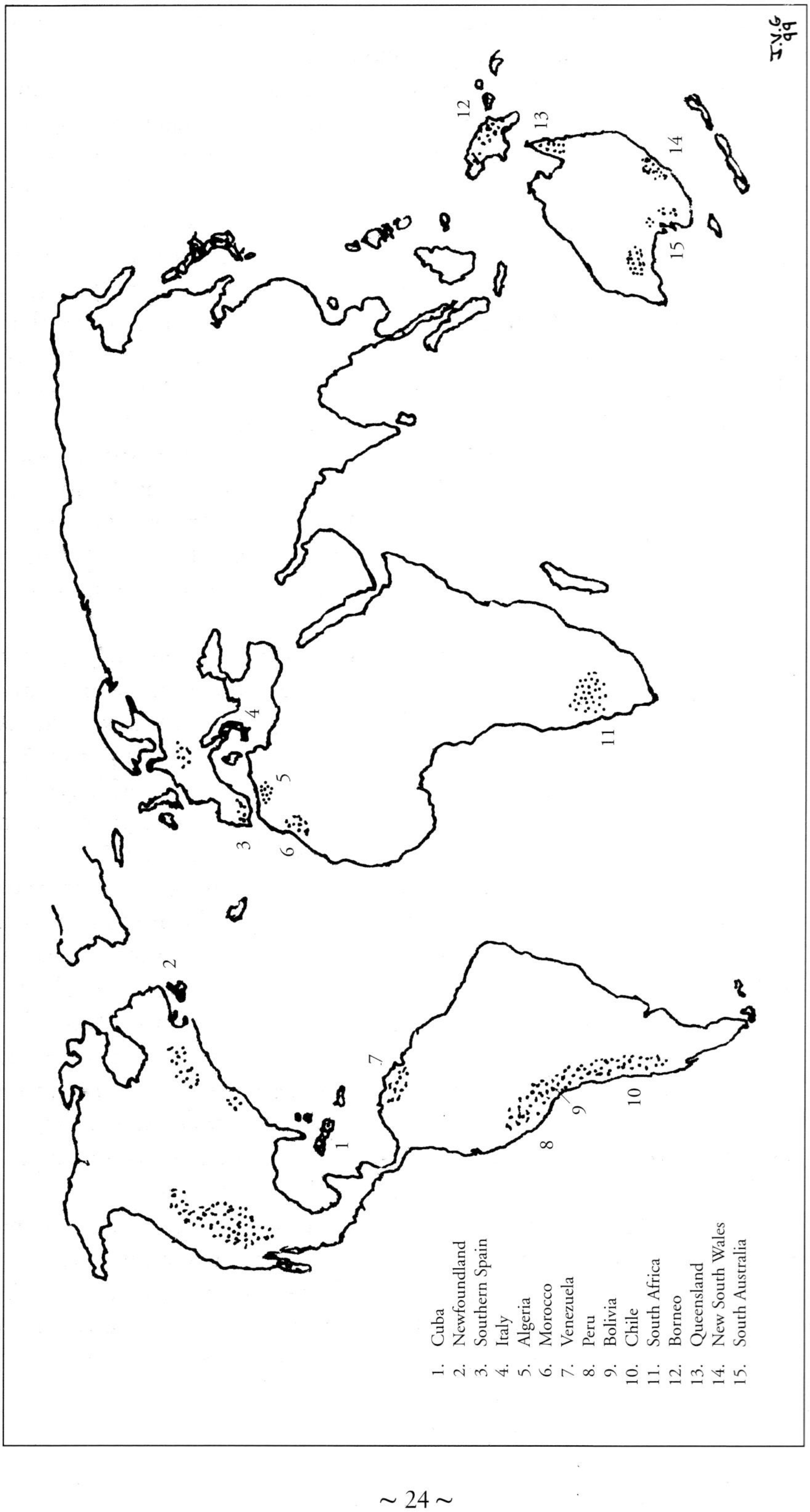

The main copper mining areas of the World.

of their period, rather than those of our times, a point which seems to be occasionally overlooked by some contemporary sociologists.

The copper men required great physical strength and endurance, and were constantly exposed to immense heat from the furnaces, perhaps ranging from 150 to 300 degrees. It meant that they sweated profusely, and their situation was not helped by the fact that the front of their bodies, facing the furnace, was subjected to high temperatures, while their backs were many degrees lower. The men were expected to work in two hour shifts, and found it necessary to consume large quantities of water. The atmosphere in which they had to work was heavily polluted, and the only means of alleviating their situation was to cover their nose and mouth with a piece of damp cloth. When conditions became unendurable they would have to rush to the door, and gulp in draughts of fresh air. Sulphur fumes affected their lungs and stomach and frequently they would spit blood.[1]

All the work was done manually; the ore was loaded into trucks which were pushed into the yards, where they were weighed into boxes of one hundred weight. The men carried these on their shoulders and emptied them into the iron hoppers which fed the furnaces. The work continued twenty-four hours a day, and when gas light became available night work was expected to be as productive as in the day.

The scene was Hogarthian, with the workers, men and women, begrimed with the yellowish green powder of the ore, toiling in the polluted air. Women and even girls were employed to help keep the furnaces supplied; 'barrow' girls used to wheel copper ore, coal and ashes to and from the furnaces, working from 6 a.m. to 2 or 3 p.m. In the *Cambrian* newspaper for 8th December 1854, Dr. Thomas Williams estimated that a copper man could 'sweat about 600 gallons in a year, and consume as much as 800 to 1,000 gallons of water to replace it.'

It is a graphic comment on the times to report that he goes on to remark: 'I would certainly be disposed to say, that the men employed in the copper works, and living almost constantly in the smoke, are not more subject to disease than in other localities, where there is a dense and crowded population, and where the individuals are constantly subject to violent alterations of temperature'. As today, one cannot help suspecting a link between the funding and reports of experts. At least it is pleasant to record that the copper workers earned more than the iron workers or miners. In the 1850s furnace men earned twenty-five shillings a week, often with housing subsidised by their employer. A foreman could expect forty-eight shillings a week, while a labourer was paid twelve shillings.

These relatively high wages recognised their special skills, as well as the hard physical conditions. For instance 'finery men' had to mix anthracite and small pieces of coal in careful proportions of four to one, in order to create a layer of clinker about twelve inches deep on the fire bars, with the same depth of coal burning on top. At regular intervals the clinker had to be broken up with great care

1. Swansea Museum Fact Sheet No. 11, page 14.

to ensure the successful operation of the furnaces. The copper owners required their workers to preserve the secrets of their trade, e.g. in 1811 William Howell bound himself to serve J. H. Vivian as a refiner for 21 years, and 'not to disclose (except to his own child or children) the Art and Mystery of Smelting or refining of Copper as practised in the Hafod Works without the consent of John Henry Vivian or his partners'.

This was extended to pressure by rival employers on their workforce not to associate socially with employees from rival firms, in case they should pass on the secrets of the various special processes. Employees of a particular firm tended to use the same public house, and in case of chapel-goers, the same chapel.

To list all the copper works in the area, even with the briefest of details, would be outside the scope of this book, but to provide a general picture, at least the main works must be mentioned. They were spread out along the coast on either side of Swansea, roughly between Llanelly and Port Talbot, drawing their supplies of ore from all over the world. The Americans, who also needed ore for smelting, used to say of the copper ore ports 'it's no good going there, a Swansea man will already have backed his cart in'.

THE PENCLAWYDD WORKS

One of the early copper smelting works in the area was established at Penclawydd in the 1780s since the records exist of copper ore coming in from Anglesey at that time. The works, which was never large, employing only about 30 men, was owned by the Cheadle Brass Wire Company, but there was also a Cornish connection in the person of John Vivian, founder of the great Swansea firm of that name. Supplies of coal were available locally, but by 1823 a combination of bad management and a diminishing local coal supply made the works uneconomic, and it was closed down.

THE FOREST COPPER WORKS

In the 19th century the Millbrook Iron and Steelworks stood just north of the present railway viaduct at Landore, and before it, on the same site, was the first copper works in Swansea, set up by Dr. Lane in 1717. John Lane was a well-known doctor and surgeon in Bristol, then the heart of the English copper manufacturing industry. He invested in copper mines in Cornwall, where his wife's step-father had connections, and then set up his works, which was named Llangyfelach after the parish in which it stood. Lane went bankrupt in 1726 and died in 1740.

Robert Morris came from Bishop's Castle in Shropshire and managed the works, settling permanently in Swansea in 1726. By 1752 Morris had set up the Forest Copper Works, which soon overtook the Llangyfelach works.

Between 1822 and 1824 Starling Benson of Penclawydd became involved with the works, which was then operated by Usbourne, Benson and Co. Starling Benson later shifted to Fairy Hill, near Reynoldston in Gower. He later became a well-known advocate for improvements in Swansea's dock and railway system.

Between 1848 and 1851 the works were owned by the Bank of England, which suggests a period of financial difficulty. In 1867 it was taken over by the Vivians, who merged it with their Spelter Works, which operated under the name of 'Birmingham' (or Ynis) works, and was still operating in the 1920s.

THE LANDORE COPPER WORKS

These were opened in 1787 by John Morris. By 1823 it was either managed or owned by the Vivians, followed later by R. J. Nevill and Henry Bath. In 1840 it passed to Williams, Foster & Co., who operated it until 1876.

THE HAFOD WORKS

Founded in 1810 by John Vivian; it became the largest in the area. The Hafod Works were described by R. O. Roberts as being 'the largest non-ferrous metal smelters in the world in the mid-19th century'.

THE ROSE WORKS

The Rose Works, built on the west bank of the Tawe at Plasmarl, was opened in 1780 by Fenton & Co. of Leeds, who were also known as the Chacewater Company because they owned copper mines at Chacewater in Cornwall. By 1824 it had been taken over by Williams, Foster and Co., who were closely associated with Pascoe Grenfell.Copper continued to be smelted at the Rose Works until 1878.

The Williams family were very much a Cornish-based concern, and John Michael Williams, who in 1840 took over as manager of Morfa, Rose and Forest, was a very forceful personality. When John Michael died in 1880 he was popularly believed to be 'the richest man in Cornwall'.

THE WHITE ROCK WORKS

John Coster was a Bristol man who perfected the art of copper smelting in the late 17th century. His son John set up the White Rock Smelting Works on the banks of the Tawe in 1737 and by 1744 it was firmly established as the first large scale smelting works in the Swansea area. In 1871 it became a lead and silver works under the ownership of Williams, Foster and Co. of Morfa, in partnership with the Vivians. By 1874 the Vivians had assumed full control, and it continued in production until the 1920s.

THE MIDDLEBANK WORKS

The Middlebank Works were established by Chauncey Townsend, a London businessman in the mid 1700s, and was situated on the banks of the Tawe above the White Rock Works. It was later taken over by the Grenfell family.

'TICKETING'

The purveyors of copper ore whether representing mining interests or ore importers conducted their transactions by a form of auctioning known as 'ticketing'. The practice began in Cornwall in the 1720s and were first held in Swansea in the old Assembly Rooms. By 1815 they had become a regular feature of life in the town, taking place at the Mackworth Hotel and preceded by an elaborate dinner, presided over by the principals of the mining and smelting companies.

The ore was divided up into parcels, and each buyer was required to place his bid, in the form of a 'ticket' on each parcel. Reserve prices were not allowed and no parcel of ore could therefore be withdrawn by the vendor.

The next stage was for all the bids to be opened, the President of the ticketing reading aloud the offers. The winning bids were published in newspapers throughout the country, and established the national and, in the 1840s and 1850s, the world price for copper.[2]

A reporter for *The Morning Chronicle* who attended an 1850 ticketing was surprised by the absence of excitement that such large scale selling and buying might be expecting to attract. The clerks sat quietly at long tables with bottles of ink, pens and blotting paper before them, writing out the tickets, which would then be folded and handed to the chairman.

The ticketing process had various effects; most importantly providing a prompt sale with guaranteed payment. However, it inevitably lent itself to the formation of 'rings' such as operate in auction rooms today, where a group of dealers get together before the sale, and agree not to bid against each other for a particular item. After the sale they auction it off among themselves and everyone gets a payoff for not driving up the bidding.

2. Swansea Museum Fact Sheet No. 7, page 16.

SOURCES OF COPPER ORE

In the 1850s, Cornwall was still the world leading producer of copper ore, and continued to rise until 1860, when it peaked at about 181,000 tons annually. In 1865 it had dropped to 160,000 tons; by 1870 it had fallen to 81,000 and in 1880 was only 40,000 tons.

In the meantime foreign sources had taken Cornwall's place. On 3rd July 1846 the *Cambrian* newspaper reported that: 'the first cargo of copper ore direct from Australia arrived on the 27th June in the brig *Amelia* of Kirkcaldy, Reed master, with 200 tons from the Burra-Burra mines'. Spanish ore averaged about 5,000 tons in the 1850s and 60s, doubling in 1870 and reached 30,000 tons by 1890.

Cuban ore was an important source; about 15,000 tons came from there in 1853, rising to over 22,000 tons in 1855. By 1860 it had dropped to just over 16,000 tons and by 1870 the Cuban copper trade had come to an end.

The most important long term source was Chile; over 13,000 tons were imported in 1853, rising to 30,000 tons in 1860. By 1870 it was down to 6,000 tons and less than 2,000 in 1880.

British South Africa was the other long term supplier; imports from there rose steadily from 1,500 tons in 1855 to 33,000 tons in 1890. But by the 1880s the copper ore trade in Swansea was in decline, basically because it was obviously more profitable to smelt the ore on the spot and export the copper as ingots. Opinions vary over which vessel brought the last cargo of copper ore into Swansea, but perhaps the most reliable could be that of Captain E. O. Evans, who was the subject of an article in *Sea Breezes* in April 1951. Captain Evans, then senior master of Furness Withy & Co Ltd., was an apprentice on the 735-ton barque *Aberlemno*, owned by Morgan Tutton of Swansea, when it returned to Swansea in 1902 with what he believed to be the last copper ore cargo brought from Chile.

COPPER-SMELTING AT NEATH

As so often, the availability of local coal and ore from Cornwall led naturally to copper-smelting in the area, which commenced as early as the late 16th century. The River Neath was tidal as far as Aberdulais, and accessible at high water to the shallow draught craft of the times.

In 1694 a smelting works was established at Cwmfelin, near Neath Abbey, but the trade began in earnest when Sir Humphrey Mackworth, the local squire and owner of Gnoll House and its estate, set up furnaces at Melincryddan in the following year, which was later connected by a specially built canal to the Neath River, half a mile away.

The Melincryddan site had the special advantage that its coal supply came from Mackworth's own collieries, and could be transported by a 'downhill' tramway, drawn by horses, directly to the furnaces. The works also produced lead, and later (by importing zinc) brass, which was used locally in the manufacture of such domestic items as kettles and pans. Sir Humphrey died in 1727 and was succeeded by his son, Herbert, who died in 1765.

Eventually the works, together with the Gnoll Estate, descended into the hands of the widow of Sir Herbert Mackworth in 1791. The following year she married the iron master Hanbury Leigh and moved away to Pontypool, leaving Gnoll House uninhabited.[3]

In the late 18th century, Roe and Co. set up a copper works on the right bank of the Clydach tributary, which after a few years became the property of the Cheadle Copper Company (who also owned the works at Penclawydd). By 1821 operations had come to an end and the works were acquired by the Neath Abbey Iron Company for shipbuilding and engineering.

The Port Tennant Canal, which linked the Vale of Neath to Swansea docks, severely affected the old port, which could only accommodate vessels of very shallow draught compared to Swansea. A works was also set up at Briton Ferry in 1853, but was dependent on supplies of ore being brought in from Swansea by rail. By 1890, as in the rest of Swansea area generally, copper-smelting in Neath was in severe decline.

3. Clive Trott, *The Copper Industry – Neath and District. A Symposium*. Edited by Ellis Jenkins.

3.

Port Development: Swansea, Llanelly and Neath

IN THE 19TH CENTURY 90% of the world's copper was smelted in the Swansea Valley, and by the time the trade reached its zenith around the year 1865 Swansea was one of the world's major ports, famous for the quality of its ships and the skill of its seamen.

In October of that year, the Prime Minister, Lord Palmerston, died two days before his eighty-first birthday. His life spanned an extraordinary period of change; when he was born, the oldest inhabitants of Swansea were alive when William and Mary were proclaimed after the revolution of 1688. Those alive at his death might have lived to hear Winston Churchill announce the surrender of Germany in 1945.

When he was born in 1784 Louis XVI and Catherine the Great ruled in France and Russia and Frederick the Great was King of Prussia. During his lifetime the population of Britain rose from about 9 to 25 million. Railways, steamships, gas lighting, the electric telegraph, and anaesthetics all arrived during that period.

In 1840 William Henry Fox Talbot invented the first practical photographic process, and in Swansea his contemporary and friend the Rev. Calvert Richard Jones, clergyman, musician, mathematician, talented oil painter and Rector of Loughor, was experimenting along similar lines. His photographs of vessels in Swansea harbour in the 1840s are probably among the best as well as the earliest photographs in the world.

The United Kingdom then ruled more than a tenth of the Earth's population, and constituted the wealthiest and most powerful country in the world. The key to this dominion by a small island in the North Sea was its mastery of the sea, and nowhere was this more evident than in the bustling, thrusting port of Swansea, with its neighbouring towns of Llanelly and Neath, and its copper smelters, ship owners, shipbuilders, merchants and seamen.

In the eighteenth century, well-to-do persons, seeking to offset the obesity, gout and dyspepsia caused by over-eating, excessive drinking and probably bad drains, frequented either spa towns like Bath, Harrogate or Tunbridge Wells, or seaside watering places such as Brighton or Weymouth. In the late 1700s, Swansea, in a

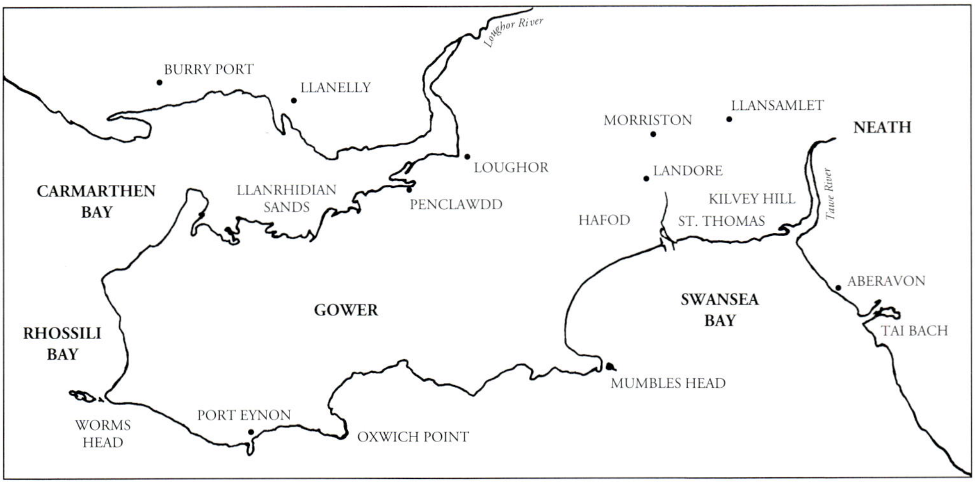

Sketch map of Swansea area.

more modest fashion, could claim to be a seaside resort, patronised by the gentry of Glamorgan and Carmarthenshire, or sometimes further afield.

There was a good coach service between Bath and Bristol, and in 1788 James Lake, proprietor of the Mackworth Arms in Wind Street, Swansea, advertised his inn in Felix Farley's *Bristol Journal* as: 'a house of truly comfortable accommodation for the nobility, gentry, and for gentlemen travelling on business'.[1]

Swansea – the crowded West Bank in the 1840s.
(National Maritime Museum, Greenwich)

1. David Boorman, *The Brighton of Wales.*

Invitation card to cutting first sod of New Dock, Swansea, 20th July, 1904.

(Courtesy Paulette Pelosi-Evans, and Mr W. J. Tanner, Swansea)

North Dock, Swansea, circa 1906.

South Dock, Swansea, circa 1890.

East Dock, Swansea, circa 1900.
(Courtesy Roy and Anne Campbell, Swansea)

From 1807 a horse-drawn coach on the Oystermouth Tramroad transformed Oystermouth with its fishing boats, oyster-fishermen, church and castle into an accessible attraction. The Mumbles Railway was the first passenger railway in the world. However, Swansea could not really hope to compete with Weymouth or Brighton, whose siren song included the presence of the Royal Family and, not least, that scandalous but entertaining ornament of high society, the Prince Regent.

Nothing could conceal the unwholesome and odoriferous foundations of Swansea's industrial prosperity on its eastern side, where the copper works, iron furnaces and coalmines that fringed the Tawe had already begun to transform the beautiful Swansea valley into a nightmarish landscape of smoke, grime, noise and smell. It was scarcely surprising therefore that hopes of a spa town were replaced by the reality of a dynamic industrial centre.

The copper-smelting works at the mouth of the Llwchwr (Loughor), Tawe and Neath rivers were maws that consumed vast quantities of coal and copper ore. At the beginning of the 19th century a ton of smelted copper required about thirty tons of coal and twelve of copper ore. The Swansea Valley was criss-crossed with the wagon roads, tramroads, and canals that carried coal from the mines to the calciners and smelters along the river estuaries which lined the coast. All the ore came in by sea. At first it was from mines in Cornwall, Cardigan, Anglesey and County Wicklow in Ireland. When Edward Donovan wanted a room at the Taibach Inn in 1804, the landlord had to explain that in a small town filled with miners, smelters, and seamen, the house would 'always be crowded with captains of vessels and their mates, whose business brings them to the copper works.'

Eventually the British mines reached a point where deeper or poorer lodes rendered them unprofitable. Surprisingly, among the first foreign ores to arrive at Hafod in 1827 came from as far afield as Australia, followed later by Spain, North Africa, Cuba, Chile and South Africa.

The facilities for loading coal and unloading ore at all the ports along the coast were crude in the extreme. The estuaries were shallow as well as tidal, and Taibach could only boast an unprotected wharf on a narrow, twisting section of the river. Penclawdd and Llanelly were even worse.

At Swansea, the mud and alluvial deposits carried out from the Tawe on the ebb tides had built up a sandbar across the mouth of the river, and only vessels with a draught shallow enough to negotiate the nine foot high bar could enter or leave at any state of the tide. Others had to wait for the flood tide or anchor in the roadstead where their cargoes were unloaded by lighter vessels until they were light enough to enter the port.

In 1768 a group of copper-smelters, ship owners and merchants presented a case to the Swansea Corporation to petition parliament for a harbour bill to improve the port facilities. The corporation at that time was dominated by a character called Gabriel Powell who was Steward to the Duke of Beaufort, owner of a large part of the town. Contemporary reports suggest that the chief preoccupation of this individual was to oppose progress wherever possible, and feather his own nest at the same time. He used his position to warn that 'strangers and meddlers' would come into the 'poor town' and by paving the way for an independent harbour authority, ultimately erode the powers of the Corporation.

His opposition was sufficient to ensure that the application was never made, and the only tangible results were various half hearted attempts to remove the banks of gravel and sand that had accumulated in the river. The subject was raised again in 1787 and once more Gabriel Powell stifled it. Feelings ran so high at the meeting that Charles Collins, an eminent surgeon and a vigorous activist for port improvement, was knocked to the floor and kicked, suffering the additional indignity of losing his wig in the process.[2]

It was not until after Powell's death in 1789 that things began to improve, and in 1791, the Harbour Act of that year set up the Harbour Trust, consisting of nominees from the corporation and mining, manufacturing and shipping interests.

In 1800 there were no large ports along the South Wales coast; all were tidal, and vessels lay on the mud when the tide was out. The creation of floating docks cost a great deal of money, which was not readily available, and only gradually did the port facilities follow the demands of local industry and shipping. Communication between towns was limited, and not helped by the nature of the hinterland, which so often took the form of valleys between high ground. Coal had to be transported laboriously down these until it came within reach of the rivers and, later on, canals.

THE NORTH DOCK

The Swansea Harbour Act of 1791 was entitled: 'An act for repairing, enlarging, and preserving the Harbour of Swansea in the County of Glamorgan', and its first board of trustees met in the same year. The initial priority was to build a dock in which vessels could float at all states of the tide, but it was not until 1852 that this was finally accomplished. In the meantime vessels of any size bringing ore from Cuba, Chile, or anywhere else, usually called first at Liverpool to discharge enough cargo to reduce their draft enough to get alongside at Swansea. This meant that the ore had to be trans-shipped to Swansea by coastal vessels, and there was a considerable coastal trade between the two ports.

As W. H. Jones remarks in his *History of the Port of Swansea*, if there was one outstanding feature of the Trustee's earliest proceedings it was the 'dilatoriness, approaching indolence, which characterised their movements'. There was a deadening frequency of abortive meetings in long succession caused by no attendance or the lack of a quorum. During the next three years the only tangible results of their

2. W. H. Jones, *History of the Port of Swansea*, page 61.

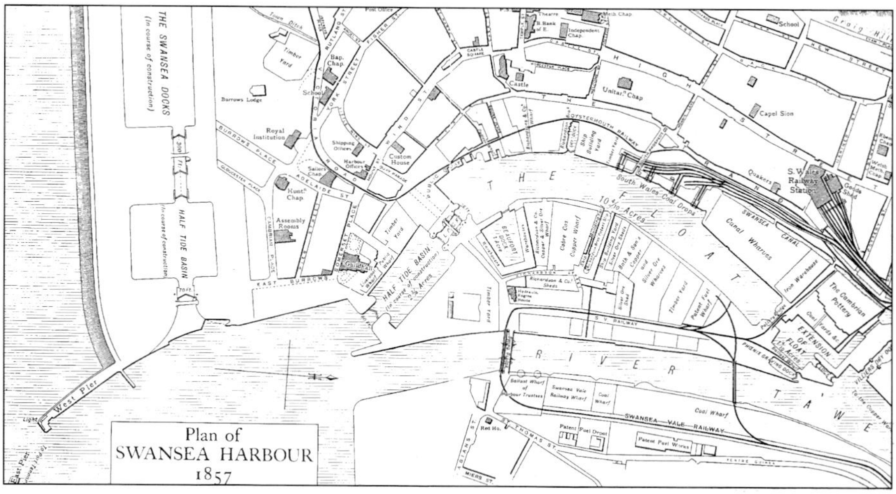

Plan of Swansea Harbour.
(Courtesy Swansea City Archives Service)

long-winded deliberations were the building of break-waters at Swansea, and the lighthouse at Mumbles.

In 1794 the trustees commissioned a report from a leading marine surveyor named Joseph Huddart. His main recommendations were ignored; the only tangible results being a number of half-hearted attempts to deepen the channel by dredging, and building piers to reduce silting. The real need, which was for a 'floating' dock, that is to say, a dock where vessels did not have to lie on the mud at low tide, was pushed into the background.

In spite of Huddart's advice, and the example of Llanelly, where a floating dock had already been created, and the vociferous complaints of the copper-smelters, ship owners, merchants, and manufacturers, ships at Swansea had to sit on the mud for another fifty years, with all the strain on their hulls that naturally ensued.

This dismal state of affairs was dramatically changed by the arrival on the scene of John Hussey Vivian, Sir John Morris, and John Richardson, and other vigorously thrusting owners of copper works, copper ore vessels, and shipbuilding yards.

In 1826/27 they commissioned the great civil engineer and road builder Thomas Telford to produce plans for a new floating dock, with upper and lower basins and a toll bridge across the Tawe. But nothing could penetrate the lethargy of their fellow port trustees, and the situation remained unchanged until 1832. Even then three more years of discussion were needed before a plan was finally submitted to parliament.

In terms of wet dock facilities, Swansea was woefully behind adjacent ports like Llanelly, where a wet dock (basically a tidal basin), was built in 1799.

In 1831, Swansea was still entirely dependent on the banks of the Tawe, with a private wharf here and there jutting out into the river.

Shipping in the river, 1846.
(Swansea Museum)

Largely through the efforts of the Vivians, Sir John Morris and the Richardsons, all this changed when at last a decision was made to create a floating dock and a New Cut, connected by suitable bridges between them and the banks on either side.

In the spring of 1840 the first sod was cut for a navigable new channel – to be named the New Cut. Work on making a float of the old channel began in 1849 and completed in 1851,[3] and the new (North) Dock appeared on New Year's day in 1852 before a crowd of eighty thousand people. The Town Float dramatically improved the facilities of the port, until then so severely limited by the tidal nature of the Tawe (see Map on page 35).

The layers of mud on which the vessels sat at low tide had begun to seriously erode with use, and consequently break down and disperse. The increasing numbers of vessels lying along the river affected the tidal flow, and created deposits of mud at the mouth of the estuary, raising the level of the Bar already existing there which vessels had to negotiate on entering the port.

As vessels lay on the mud at low tide they created depressions which corresponded to their hull shape. Vessels subsequently laying on or near the same spot might very well have had hull dimensions which did not fit the particular depression or

3. *Cambrian* newspaper, 1st May 1840.

its edge, which could place a considerable strain on the frames and planking of a wooden vessel, especially in the case of older craft. In addition, plans for the new dock had been complicated by a conflict of interests between merchants handling coal and copper ore. The former tended to use smaller vessels, of say 75 to 125 tons, which could lie on the mud without being susceptible to so much strain, so they were not very enthusiastic over contributing towards the high costs involved in building a wet dock.

THE SOUTH DOCK

The next major development was the South Dock, again largely due to the efforts of the copper-smelters headed by the Vivians. The South Dock represented a new concept in that it took the form of an excavated dock which provided a separate addition to the existing port facilities. It would cater for vessels of deeper draught and could be connected to the then expanding railway system. And of course, being adjacent to the river mouth, docking and departure lines were speeded up. The original plan was to finance the project privately in the form of a Dock Company, whose prospectus attracted considerable support from prospective public share holders. However, it soon became apparent that the cost of the scheme was beyond the resources of a private company, and consequently the Swansea Harbour Act of July 1857 brought it under the charge of the Swansea Corporation. The new dock was formally opened in September 1859.

Opening of South Dock, 1859.
(Courtesy of Swansea Museum)

South Dock, May 1888.
(Swansea Maritime Museum)

THE EAST SIDE AND THE PRINCE OF WALES DOCK

George Tennant of Cadoxton Lodge, Neath, born in 1765, was a private gentleman of means who had become interested in canals and the contribution they could make to the industrial prosperity of the Neath Valley and also to the port of Swansea. There was already a canal system at Neath operated by the Neath Canal Company, and in 1822 Tennant suggested a new canal that could join up with theirs, and connect it with the port of Swansea. A contract was drawn up, and an aqueduct built to bridge the Neath river. The new canal system was completed in May 1824, and inaugurated by the successful transit by ten coal barges from Protheroe's Colliery in the Vale of Neath to deliver their cargo at Swansea harbour.

Finally, it was recognised that a further large dock was required to meet the Port of Swansea's ever growing needs, which would connect St. Fabian's Bay and most of Port Tennant itself to Swansea Harbour. The whole system could be interlinked by the Great Western and Swansea Vale Railways, complete with extensive new quaysides, coal-drops, and warehouses.

Work on the new dock, named after the Prince of Wales, commenced in 1879 and was completed by October 1881, when it was formally opened by the Prince and Princess Alexandra, later to become Edward VII and Queen Alexandra.

Between 1881 and 1889 there was a decrease in the actual numbers of vessels using the port, but this was counteracted by a substantial increase in individual

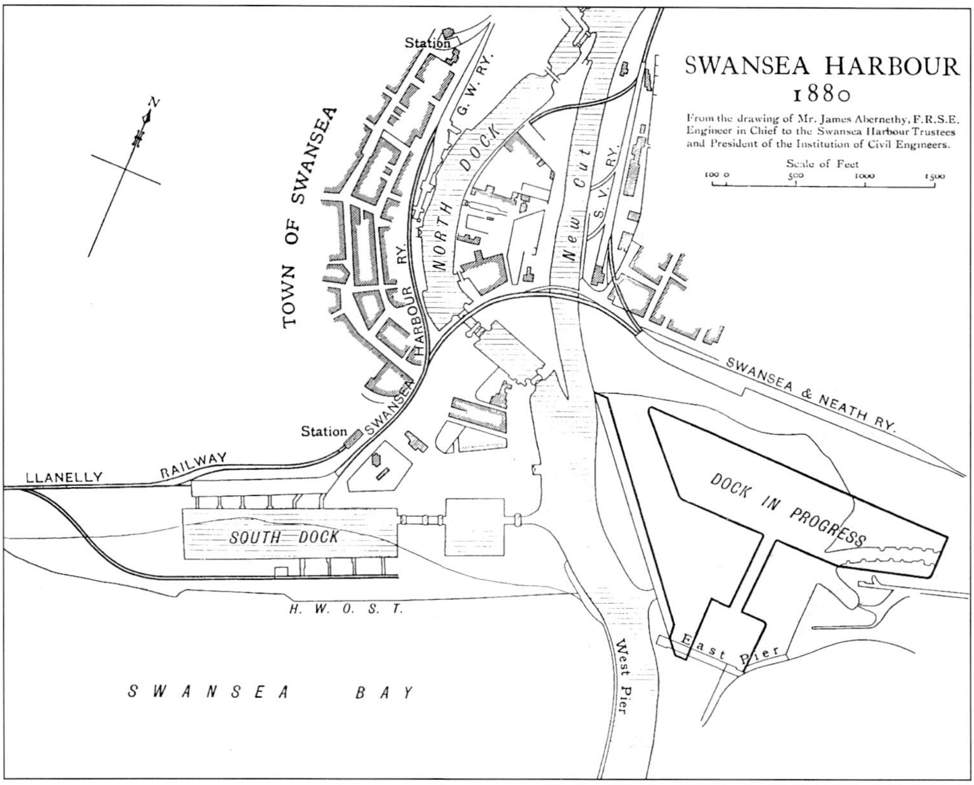

Swansea Harbour, 1880.
(Courtesy Swansea City Archives Service)

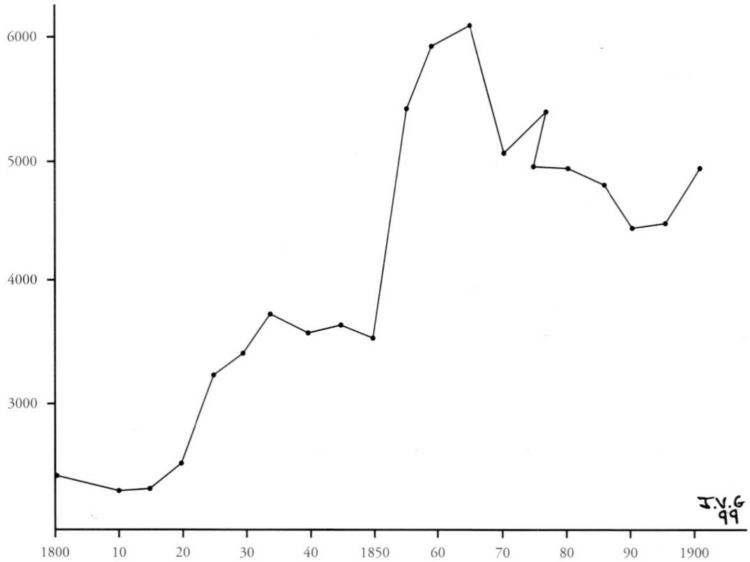

Annual number of vessels entering Swansea Harbour between 1800–1900.

tonnages. By the 1880s, steamships had overtaken sail, and the Prince of Wales Dock demonstrated in the most practical way the importance of an integrated dock system, well served by railways.

LLANELLY

As at Swansea and Neath, the presence of local coal mines and river estuaries combined to promote the smelting of copper ore imported from Cornwall. It is recorded that in 1707 more than 300 vessels left Llanelly loaded with coal cargoes.

The River Loughor flows into the Burry Estuary which forms the boundary between Glamorgan and Carmarthenshire, is about six miles across at high water. With its winding channels and treacherous sandbanks it was described in 1813 as 'an unsafe estuary with a limited coal trade, suitable only for vessels below 100 tons'. The loading sites had always been primitive, exposed and dangerous. One of the busiest of these was known as the Flats, which had no shelter from the prevailing westerly winds, and the stormy tides that scoured the estuary. When loading or unloading, vessels had to lie on the mud at low tide, when they were accessible by horse and cart for short periods at a time.

In 1795 a small wharf was built by Messrs. Bowen and Roderick, which was connected by a canal to their colliery at Wern. A similar wharf was built in 1799 by Alexander Raby, which became known as Doc y Squire. When copper-smelting began in the area in 1804, the ore had to be brought in by the same system. In 1805 there was a dramatic improvement in the form of the Copper Works Dock. In July 1813, a Harbour Act was passed for 'the improvement of navigation of the rivers Burry, Loughor and Lliedi in the counties of Carmarthen and Glamorgan.'

The New Dock was opened in 1834 and the channel connecting it to the Flats was improved but still the port was severely restricted by lack of depth in the main channel. In 1843 a second Harbour Act led to further substantial improvements, including a reservoir whose function was to scour the harbour. But the biggest improvement arose from dredging and widening the six mile channel that lay between the bar and the town. From this period onward the various industries of Llanelly which owed their existence to coal and the port began to mushroom.

In 1840 Garret was building vessels at Joblings Wharf and anthracite was being exported. (Prior to 1800 the chief port in South Wales for the export of anthracite was Milford Haven.)

The Llanelly Iron Shipbuilding Company was formed in 1863 by W. H. Nevill of Llanelly, whose family had been building ships at Llanelly for more than forty years. The yard was near the Carmarthenshire Dock, and the first vessel to be launched there was the iron brig *Premier*, 220 tons, in July 1863.

One well-known vessel, the *Hinda*, 440 tons, owned by Simon Goldberg of Swansea was launched in 1869. Others included the iron schooner *Jamapa*, 195 tons in

1873, and the iron brigantine *Grijalva* in 1874, both owned by Thomas Harrison of Liverpool.

Building continued until the late 1870s, when the yard was taken over by Samuel Brothers, who were local shipowners. Thereafter it became mostly engaged in repair work, and finally closed before World War I.

Buckley's Brewery produced in vast quantities a beverage proudly described by the locals as 'the wine of the country'.

By the 1880s the copper trade began to diminish, but its place was more than taken up by tinplate. In 1880 there were 31 tinplate mills in Llanelly alone; 75% of their output went to the USA, and the trade continued into the 1930s, when severe competition from abroad drove it into decline.

I cannot leave the port of Llanelly without referring to the terrible calamity that occurred in January 1868:

> "About twenty vessels which had been wind-bound in the port left on a Wednesday evening tide towed by steam tugs. On nearing the bar, the tow rope snapped. Sail was made with all at speed, but the wind having lulled and the breakers on the bar being heavy in consequence of the recent gale, the vessels were not able to get clear of the heavy sea that came rolling in upon the shore. Night closed on the scene and those who had witnessed the departure of the vessels hoped they would get clear of the dangerous breakers. This hope was not realised; the ships could not beat out to sea and finally they drifted upon the sands. The survivors say a more fearful night they never passed."

Eight vessels were totally lost, and next morning Llanelly 'was a town of grief and desolation'. Some survivors swam to Pembrey, others to Gower, but most never reached land. The *Mary Fanny* ended up high and dry 'some distance beyond Llanrhidian, but the captain and crew had perished like so many others on that terrible night.'[4]

4. Gareth Hughes, *A Llanelly Chronicle*, pages 100-101.

4.

The Copper Smelters, Shipowners and Shipbuilders

AT FIRST SIGHT IT MAY appear both unwieldy and confusing to include in one chapter a description of three quite separate activities. However, any other approach could only lead to an uncoordinated narrative. The owners of copper-smelting works were frequently shipowners; shipbuilders were sometimes copper factors and shipowners, and all three tended to have fingers in each other's business pies.

Inter-marriage between the great mercantile families of Swansea added to the richness of the mixture, and all came together in debating, and contributing to, the Corporation of Swansea's development of the port and its facilities throughout the Victorian era.

In describing individual shipowners, of whatever category, the names and details of the vessels they operated must clearly be an important ingredient. But simple lists are misleading. In the Victorian era, the ownership of a vessel was frequently divided up into blocks of sixty-fourth shares, so that the names of several owners can appear against a single vessel. The system had many advantages, the most important probably being the sharing of insurance risk, but it also enabled members of the local business community, at all levels, to procure an active investment in an industry which might be directly or indirectly connected with their own business.

Thus some vessels were owned outright by one individual; others were shared. Some were chartered from outside Swansea, or between each other. Ships were bought or sold, or passed between members of the same family, or families linked by marriage. Holding companies were created, such as the Swansea Merchant Ship Owners Company Ltd., only to be wound up a few years later. The situation can best be imagined as a sort of continuing vortex of thrusting Victorian entrepreneurship.

The copper barques alone numbered many hundreds, and this book could easily be expanded by long lists of vessels, some with exhaustive details, others with little more than a name and the most basic of details and as boring as a train-spotter's notebook. I have therefore adopted the compromise of selecting the most prominent owners and attaching them to a list of the vessels most clearly known to be associated with them, together, when available, details of particular interest. These appear as Appendix I and II.

THE BATH FAMILY

Henry Bath (January 1776 – May 1844) was a Cornishman who in 1794 founded a copper-trading business in Falmouth which inevitably developed strong links with Swansea, then the rapidly growing centre of the UK copper smelting industry. He was a descendant of Sir Henry Bath, Knight, who flourished during the reign of Henry III, and appears in John Prince's book, *The Worthies of Devon*, published in 1697. It describes the family as 'of great antiquity' and mentions a Sir Walter de Bath, who was Sheriff of Devon in the last year of the reign of Henry III, 1217 AD.

He was born in Falmouth, 24th January 1776, and on 7th November 1796 married Elizabeth Paddy, by whom he had two sons, Henry and Edward, and two daughters. In 1799, like his father and grandfather before him, Henry became a member of the Society of Friends, and moved to Portreath, Cornwall. In 1820 the family moved to Swansea, where he built himself a house at Rosehill, Oystermouth.[1]

A minute of Swansea Corporation records his acquisition of a 'wharf and premises' on 15th January 1822 which became known as 'Bath's Copper Ore Yard'. It was situated on the riverside to the north of the Ferry House, and the lease was for sixty years at an annual rent of £41 per annum plus port dues, dating from the 25th March 1821.[2]

At the rear of the site was the terminus of the Oystermouth tramline at Bath Lane, so named after Henry Bath. There was a kiln situated there where limestone from Gower was burnt, which under the terms of the lease had to be destructed and never rebuilt. By 1822 he had moved the headquarters of his business from Falmouth to Swansea, and taken his son, Henry Bath II, into partnership. Bath and Son owned and operated a fleet of vessels, at the beginning engaged in the UK coastal trade, and later internationally, running to Spain, North Africa, Cuba, and finally Chile. The Bath operation covered all aspects of the copper trade, including sampling the ore on arrival, weighing it, setting freights and landing charges, and finally arranging for its sale by the ticketing process.

Henry Bath II was born at Falmouth in August 1797, and came to Swansea with his parents in 1820. He married Susan Madge, born 2nd September 1798, died in January 1861. Henry II built himself a town house, Longlands House, on the site of the present YMCA building at the bottom of Page Street, Swansea, then on the outskirts of the town. He died in Falmouth in his 68th year, and his body was

1. Swansea City Archives Service. The Bath Family Records.
2. *ibid.*, T.C. 100/Index 15. Report of the Estate and Property of the Corporation of Swansea, 1850, page 5.

A Swansea shipyard, circa 1840s.
(Swansea Museum)

brought back to Swansea for burial in the Quaker Cemetery in High Street, not far from where the railway station now stands. In Victorian times the Bath family were Quakers, and a memorial stone (now standing against a wall) bears the name of several members of the family, including Henry Bath I, who died in May 1844, and his mother, who died in March 1820, aged 79.

He had three sons, Henry James, Edward, and Charles. In the late 1840s Henry Bath II established the Swansea Iron Shipbuilding Company, and in 1849 launched a steam-yacht, the *Firefly*, which was intended for use on the coast of Chile. She arrived safely at Coquimbo after a sixty-four day passage around Cape Horn.

Henry James and Edward both went out to Chile in the course of the firm's business, and while there they married the two daughters of Charles Lambert of Chile and Swansea, another shipowner and copper merchant and smelter. The Lambert and Bath families subsequently became closely associated in business, Edward managing his father-in-law's copper smelting works at Port Tennant, as well as being involved with Henry James in running the various Bath enterprises. The Lamberts had by then set up a copper smelting works in Coquimbo, their example being subsequently followed there by other UK firms. Edward Bath lived at Brynymor, and died suddenly while addressing a council meeting at Swansea on 23rd October 1885.

Throughout the 60s the Baths had vessels building at Bideford in North Devon. Their vessels were named after the letters of the Greek alphabet, Z being the barque *Zeta*, which was fitted with an auxiliary engine. At the time this was a considerable innovation, but the experiment was not successful. After a few years she was sold on the Chilean coast because the cargo capacity taken up by the engine and the coal it required adversely affected her commercial viability.

Other vessels' names were inspired by the novels of Fennimore Cooper, set among Red Indians in the forests of Canada. These included the *Scout*, *Hawkeye*, *Uncas*, *Delaware*, *Pathfinder*, *Mohican* and *Deerslayer*.

The figurehead of the *Pathfinder* aroused particular admiration; the sculptor had faithfully represented his fringed buckskin leggings, embroidered shirt, moccasins and, in the best Davy Crocket tradition, his coon-skin cap. Hanging from his waist was a powder-horn, hunting knife, shot-pouch and water-bottle, and while holding a rifle in his right hand he leant forward, shading his eyes with his left hand. It would be hard to conceive a more fitting subject for the figurehead of a vessel whose career was to be spent thrusting her way through the southern seas to Chile and back, doubling Cape Horn each time in the process.

Henry James and Charles carried on managing the fleet and their copper yards in Swansea, and during the 1850s they opened an office in London. Transactions were at first carried out there rather informally in various trading offices, and frequently over coffee, which was then all the rage, in the Jerusalem Coffee House, just off Cornhill. In January 1897 the London Metal Exchange Company was formed to provide a more regulated forum for their transactions, and its first venue was in a room over Christie's, the makers of hats, whose premises were situated at 4, Lombard Lane.

Henry James Bath was born on 17th December 1821; he was educated at the Society of Friend's school at Sidcot, and joined his father in the business when he was 18. Like his brother Edward, he was closely associated with the Lamberts; the two families' business interests ran side by side, both in Swansea and Chile. In 1868 he built a house, Alltyferin, near Llanegwad in Carmarthenshire which he occupied until his death on 15th September 1875. He died shortly after his return from Chile, having suffered a stroke on the homeward passage. He was a Justice of the Peace for both Carmarthenshire and Glamorgan, and he inherited the house Rosehill, that his father Henry Bath I had built at Oystermouth.

From 1859 to 1872 he was a member of the Swansea Harbour Trust, and was High Sheriff of Carmarthenshire in 1869. In that year he obtained a Grant of Arms, similar to that used by the Bath family prior to their adopting the Quaker faith. Henry James was a staunch member of the Church of England and built a chapel on the Alltyferin estate. He was buried in Llanegwad church. His widow lived on at Alltyferin until 1885, dying in Oxford in 1902. There were no children to succeed them.

Charles Bath, the third son of Henry Bath I, was born in Swansea, 15th January 1832, and lived at Fynnone, Swansea. He was a J.P. for Glamorgan and Swansea; F.S.A., Consul for Italy, and Vice-consul for Brazil. He became Mayor of Swansea in 1864. His wife was Emily, born in August 1835, who died 15th December 1896, whom he married in August 1856. He died 26th December 1888, leaving no issue.

With the opening of the London Exchange the firm began to take on an increasingly widening international aspect. By 1875 an office had also been opened in Liverpool, along with associated metal warehouses. By that time the Baths had also become involved with the Chilean nitrate trade, a connection that continued until Chile nationalised its nitrate industry in the 1920s.

After the Second World War the company diversified into steel buying and selling. The company continued as a privately owned family firm until the early 1970s, when it was bought out by the British Bank of Commerce in Glasgow. In 1975 the company became an important member of a major Australian mining group, MIM Holdings, and remains today an important member of an expanding group of companies dealing with mining and metals on a world-wide basis. One wonders what the first Henry Bath would have to say about it all, could he return and survey the consequences of his own enterprise.

THE LAMBERTS

The Lambert and Bath families were closely connected in their business activities, and by three inter-marriages between 1848 and 1852. Charles Lambert, who became an important figure in the copper industry, both in Swansea and Chile, was the son of Joseph Lambert, a Justice of the Peace in Alsace and Depute for Bas-Rhin. He was born at Lauterbourg, 31st December 1793, and was educated at the Ecole Polytechnique in Paris, where he trained to become a mining engineer.

After qualifying he went out to Chile to work in the mining industry, where he was employed for about four years before setting up his own smelting works at Coquimbo and La Serena. About 1820 he married Janet Spears, a Scotswoman from a Kirkcaldy family.[3]

In 1840, he settled in Swansea, with his numerous family, having become acquainted with the Bath family as a result of an introduction from Mr. William Gibbs of Gibbs & Co, of Chile. The senior partner of that firm was Mr. Henry Hucks Gibbs, a Governor of the Bank of England, and an M.P. for the City of London. In 1897, he accepted a peerage under the title of Lord Aldenham.

3. *ibid.*, The Lambert Family.

In 1852 Charles Lambert acquired land at Port Tennant, near Swansea, adjacent to the Port Tennant canal, from the Duke of Beaufort and the Earl of Jersey. There he established the Port Tennant Copper Works, near to the east pier of Swansea Docks. The works, which soon became known as Lambert's, consisted of buildings running alongside the two arms of the Port Tennant canal, which prior to the Swansea and Neath railway line of 1863 and the building of the Prince of Wales Dock in 1881, provided a sort of harbour facilities. Charles died at Alltyferin in August 1876.

His son, Charles Joseph (June 1826 – July 1888) developed the family business to its full potential and its success became a source of irritation to the owners of Swansea copper smelting works. The Lamberts' pioneering activity in smelting copper in Chile was seen as reducing Swansea to merely a finishing centre, and it was rightly regarded as an ominous portent for the future.

The Swansea works also produced arsenic, but by 1875 its advertising material was mainly concerned with sheeting, nails and bolts. The works were dismantled as the land it stood on was required for building railway sidings when the Kings Dock was built in 1904. It was a good time for the Lamberts to sell the works, for by then the boom times of the copper smelting industry in Swansea had already come to an end.

A row of workman's cottages associated with the works, known locally as 'Lambert's Row' continued to exist long after the works closed down. The Lambert advertisement in *Industries of South Wale*s (Swansea Museum collection 660) is interesting as a typical example of Victorian entrepreneurship:

> Charles Lambert & Co., Port Tennant Copper Works, near Swansea. – Port Tennant is one of the principal seats of the copper industry in South Wales, and the Port Tennant Copper Works, under the proprietorship of Messrs. Charles Lambert & Co. is one of the best known and most reliable concerns of its character in the kingdom. It was founded thirty-six years ago, and has steadily and rapidly increased in importance from the commencement, a result that abundantly testifies to the superior and uniform quality of the goods sent out, and the persistent energy and enterprise and great skill displayed in the management. The works, with yard attached, occupy twelve acres of ground, and are situated in close proximity to the East Docks, besides having direct canal communication, also railway connection with the Great Western Railway, London and North Western Railway, Midland and Rhondda Bay Railway. Thus it will be seen that splendid transport facilities are possessed, a feature of paramount importance in a business of this nature. The premises throughout are very substantially constructed, skilfully arranged, and equipped with the most

> improved machinery and appliances for the prosecution of the various branches of trade in hand. At the present time the firm employ about three hundred and fifty hands, and this by no means represents the full capacity of the works, for even with a firm of this standing the depression in trade is sure to be felt to some extent. Each department is under talented supervision, and indeed the establishment is organised throughout in a manner to ensure the prompt execution of all orders. Messrs. Lambert & Co. are copper smelters and rollers, sulphate of copper manufacturers, electro depositors, copper, silver and gold smelters and refiners, and colliery owners; and their productions in each of these branches are acknowledged to be unsurpassed in either home or foreign markets. The members of the firm, Messrs. J. A. Weir and W. L. P. Mark, London, and C. L. Bath. Swansea, are gentlemen of thorough practical acquaintance with the whole routine of the business, and by personally directing all operations guarantee the reliability of all goods sent out. They are well-known in leading industrial circles of South Wales, and greatly esteemed for the valuable part they have taken in the development of so important an industry.

Charles Lambert had two sons, Charles Joseph, Robert Spears, and daughters Margaret, Eugenie and Helen. Charles Joseph married Susan Bath and produced ten children, Charles Lambert, Catherine Susan, Janet Spears, Henry Bath, Robert Spears, Margaret Spears, Harold Penrose, Edgar Penrose, Alfred Penrose and Penrose. All of Charles Joseph's children were born in Malaga, except Janet and Penrose, who were born in London. Details of all of them are hardly relevant, but one, Charles Lambert, is a melancholy exception. He was born on 5th July 1850 and educated at Eton and Christchurch, Oxford. He went to Chile on the firm's business and during the trip was invited by Captain Cator of *HMS Scout* to visit the Sandwich Islands, where he lost his life by drowning while surf bathing at Kalua, Hawaii, on 20th November 1874.

THE VIVIANS AND HAFOD

The Vivian family, who with the Baths were the founders of Swansea's preeminence as a 19th century copper boom-town, will always be associated with Hafod, where they made their money, and their estate at Singleton, on the west and pollution free side of Swansea, where, as one might say, they spent it.

The sprawling Hafod works, which by the 1840s was the largest of its kind in the world, and their home, Singleton Abbey, were two sides of the same coin. It would be difficult to find a more perfect example of the Victorian industrial entrepreneur who destroyed the local environment and created working conditions which shortened the lives of all the workers it touched, yet at the same time provided desperately needed employment. By the standards of the day the Vivians were philanthropic employers who tried to look after their work force. And they left behind them public parks, an art gallery, a library, a University and a hospital.

John Henry Vivian provided houses for his workers; in terraced rows two up and two down, at a rent of 1/6d per week. Larger ones had three bedrooms with a bigger parlour, and a small garden at the back with room for a pig-sty, some chickens, a coal-shed, and a thunder-box. Coal was provided either free or at subsidised rates. In 1847 the Vivians built a school for 500 children; admission was free for the families of employees, and 2d per week for others. Unlike most employers they paid their workers on Friday so that they could shop in the market on Saturday morning.

If the wage-earner became ill his rent was waived, and widows were left in their houses, or at least helped to find accommodation elsewhere. When a worker reached a stage where either through age or ill-health he could not do hard work, then a lighter task was found for him.

In Welsh the name Hafod means a house in the hills, and so it was, for the landscape consisted of green-clad hills rolling down to a farm, meadows, and groves of oak trees. Scenic destruction had already begun in the late 19th century when ships using the Tawe river began to dump their ballast on land belonging to the farm, a process which created an eye-sore, and threatened to narrow the river.

The first of the Vivians to settle in Swansea was John Vivian (1750-1826) a Cornishman from Truro. He was the chief negotiator for the Cornish Mining Company, which was a consortium of mine-owners, and as such was continually in touch with the Swansea ore-smelters. By 1800 he had become interested in The Cheadle copper works at Penclawdd and Loughor on the north coast of Gower, and in 1810 he leased land at Hafod from the Duke of Beaufort and Lord Jersey in the name of his sons, John Henry and Richard Hussey. This site, which became the Hafod Copper Smelting Works, was very conveniently situated, with the River Tawe on one side and the Swansea Canal on the other.

In 1819, when John Henry was 31 years old, he set up the first metal rolling mill in Swansea, and by 1823 a 'nail manufacturing' department was in production. 'V & S' nails are still in the roofs of Singleton Abbey and St. Paul's Church, Sketty.

Together with other owners of copper works in the area, John Henry received numerous legal protests over the establishment of his works from local landowners who fully anticipated the devastating effect on the environment his activities were bound to produce. A good example of one of these was registered in 1809 by Rowland Prichard:

> "To John Vivian Esq.: I hereby give you notice that if you erect any Works at or near to a certain place heretofore called and known by

John Henry Vivian.
(Courtesy Swansea Museum)

> the nature of The Brick Yard, within the parish of St. John's, in the County of Glamorgan, I shall commence an Action against you for the injury that may be done to my Estate, situated in the same parish, in the occupation of Morgan John, by the said Works in the Smelting or Manufacturing of Copper. Dated the 22nd July 1809 – yours etc., Rowland Prichard."

Alas, for Mr. Prichard and others like him, their protests had little success; a combination of legal procrastination and the support of interested parties associated with the Swansea Corporation easily won the day.

Against that must be set the efforts of some of the copper works owners, and especially the Vivians, to overcome or at least reduce the environmental ravages resulting from their operations. John Henry Vivian had received a mining training in Germany, then at the forefront of metallurgical science, and when the Hafod works were set up in 1810 he brought over a German chemist by the name of G. B. Harman to set up a laboratory. One of its main tasks was to explore possible means of alleviating the havoc created by copper smoke. By 1812 Harman had been appointed a managing partner of the works, and he used his authority to offer a reward of £1,000 for a successful solution.[4]

One obvious way was to build very tall chimneys, but this only provided a partial solution; a real breakthrough did not occur until 1868 when a new smelting process invented by another German, Moritz Gorsten Hoffer, produced a real reduction in copper pollution and incidentally, a useful by-product, sulphuric acid.

By the 1830s, Vivian and Sons had a fleet of schooners and smacks which ran between Swansea, Liverpool, London and Rouen. These included the *Brothers* and *Henry*, both schooners of 86 tons, and two smacks, *Sarah* and *Eliza*, 46 and 56 tons respectively. By the 50s and 60s the Vivians were operating larger vessels, running to France, Spain, Cuba and Chile. Although the Vivians were shipowners, they did not themselves operate large numbers of vessels, as did for instance, the Baths, Richardsons, Burgesses, and other Swansea shipowning families. But at the same time their financial interest, at the capital level, was substantial. As the Victorian era wore on, and the Vivian's business empire grew, so did the number of rows of terraced houses at Hafod built for their workforce. In the 1840s cholera swept through Swansea, and Hafod, with its concentrated population was particularly badly hit. Poor water supplies were a contributing factor, and the Vivian's attempts to improve matters by sinking wells did not prove successful. John Henry Vivian was a Member of Parliament for Swansea during the period 1822-55, and was made an F.R.S., following a paper he wrote for the Society on the subject of copper smelting. His elder brother, Richard Hussey, entered the Army, where he distinguished himself during the Peninsular War. On retirement he was created Lord Vivian of Glyn.

Henry Hussey Vivian, John Henry's eldest son, managed the Hafod Works and assumed full control of the business when his father died in 1855 at the age of 76. He was born at Singleton Abbey, the family home in west Swansea, and was educated at Eton. Like his father before him he studied metallurgy in Germany and France. His initial job in the family business had been to manage the Liverpool branch of the firm, which he continued to do until coming to Swansea in 1845 to take over the Hafod Works. Under his management the Vivian business empire steadily expanded; there was scarcely an important mining area in the world where they did not have an agent acting for them – New York, Chicago, East and West Canada, California, Chile, to name only the most important. Even to list them would occupy more space than is here available.

Henry Hussey was created a Baronet in May 1882, and in June the following year he was ennobled as the first Baron Swansea. He was a Justice of the Peace and a Deputy Lieutenant for Glamorgan, and died in November 1894 at the age of 73.

By that time the family seat had been firmly established at Singleton, to the west of Swansea. In the early 1800s the land now occupied by Singleton Park, Brynmill

4. N. L. Thomas, *Swansea's Districts and Villages.*

Henry Hussey Vivian.
(Courtesy Swansea Museum)

Park and Swansea University was an idyllic area in excess of 250 acres, taken up by a number of small farms with mixed arable and pasture. In 1816 John Henry Vivian purchased a small farm with an octagonal white house called the Marino. The direction of the prevailing winds made it the right side of Swansea to avoid the noxious fumes emitted by the Hafod works, and provided a suitable site to build a house commensurate with his growing wealth.

He began to add to the building, which commenced to grow, rather in the manner of Topsy. By 1837 it had been transformed into an extensive mansion, sprouting in all directions the pseudo-Gothic embellishments so beloved as status symbols by successful Victorian businessmen.

It was re-named Singleton Abbey and provided a suitable background for the social activities which accompanied the position in society occupied by the Vivians. In due course it descended to his son Henry Hussey Vivian, who entertained there such figures as the Prince and Princess of Wales, Lord Palmerston, and the Earl of Cardigan. By 1919 the head of the family, the second Lord of Swansea, had moved away from the area and the Singleton estate was purchased by the Swansea Borough Council who utilised the land to create the present public parks of Singleton and Brynmill, and in due course, the University of Swansea, and Singleton Hospital.[5]

5. Bernard Morris, *The Houses of Singleton.*

THE GRENFELL FAMILY

Pascoe St. Leger Grenfell.
(Courtesy Swansea Museum)

The Baths and the Vivians, as well as devoting their lives to making money out of copper, were also, by the standards of the day, good employers who cared for their workforce. But the Grenfells, equally ruthless in their business activities, lived close to their workforce on the East Side of Swansea, and were enthusiastic social philanthropists. Chauncey Townsend, a London merchant who settled in Swansea in 1750, invested heavily in collieries in the Llansamlet area. In 1755, seeing in metal smelting a good outlet for the coal produced by his collieries, he set up, in association with London brass and copper manufacturers and his two sons, James and Joseph, the Middle Bank Copper Works between Foxhole and Pentrechwyth. By the end of the 18th century the works had become the property of an Anglesey lawyer, Thomas Williams, who later became known as the 'Copper King' because of his briefly successful attempt to dominate the copper ore and smelting industry.

Pascoe Grenfell (1761-1838), who came from St. Just in Cornwall, was also involved. It is known that he spoke both French and Dutch, and he set himself up in London as a merchant banker, his assets consisting of 'a little capital, a pipe of good wine, good experience, and Cornish connections'. By the 1780s he had acquired considerable wealth and an interest in the Middle and Upper Bank Works. By 1794 he owned a two-thirds share in both works. In 1803, in partnership with Owen Williams, son of the 'Copper King', he took over both works completely.[6]

By 1825 the Williams family had dropped out, and thereafter Pascoe Grenfell & Sons dominated the industry on the east side of Swansea until the 1890s. The first member of the Grenfell family to settle in Swansea was Pascoe Grenfell's son, Riversdale William, who in 1835 leased the Maesteg estate from a Mrs. M. E. Tennant.

He was followed in 1840 by his elder brother, Pascoe St. Leger, who built Maesteg House in the 1840s, and lived there until his death in 1879. Pascoe St Leger believed in the hands-on approach; he managed the business himself, including the vessels owned by the company, and between 1828 and 1851 a copper smelting works at

6. Muriel Chamberlain, *The Grenfells of Kilvey* (*Gower* magazine).

Holywell in Flintshire. He was a member of the Swansea Corporation, Chairman of the Swansea Harbour Trust, and of the Swansea Vale Railway. He was also a notable yachtsman, though on rather the grand scale, for he simply used one of his own vessels, a clipper ship called the *Taplow* for cruising with his family.

His sisters Charlotte and Fanny married the writers James Anthony Froude and Charles Kingsley. His home, Maesteg House, was used to house Belgian refugees during World War I, and it was demolished not long after. The present St. Thomas Community Centre now stands on its site. Charles Pascoe Grenfell, the senior partner in Pascoe, Grenfell and Sons, who had long been a key figure in the family business, died in 1867, and the family connection was severely weakened by the death of Pascoe St. Leger in 1879.

In 1890 the business became a limited company, but as all the shares were bought by the Grenfells its basic characteristic was not changed. The family links with Swansea were never the same after the death of Pascoe St. Leger, and in 1892, Pascoe, Grenfell & Sons went into liquidation. By then the great days of the Swansea copper industry had come to an end.

The philanthropic activities of the Grenfells deserve recording; Pascoe Grenfell was a friend of William Wilberforce, and frequently spoke against slavery in the House of Commons. Madelina, his eldest daughter, endowed the eye department at the Swansea General Hospital. Above all, in the style of the Baths and Vivians, they built houses for their workers on Kilvey Hill which they rented out at nominal rates – 6d per room per week for a house of the two up and two down variety. In 1842 they built All Saints Church at Kilvey on land donated by Freeman & Company, who owned the White Rock Works, and schools which could accommodate over a thousand children. Of the great Swansea industrial magnates, they were the only ones who continued to live on the east side, close to their workers.

The sudden closure of the Middle Bank and Upper Bank Copper Works in 1892 caused consternation in Swansea, for the Grenfells were considered 'one of the soundest and most lasting commercial organisations in the district'. However, the situation was not totally disastrous, for the works were bought by Williams, Foster & Co, and thereafter became Williams, Foster & Pascoe Grenfell.

Pascoe St. Leger had five daughters, and one of them, Mary, is specially remembered for her charitable work. When her brother Francis (later Field Marshal) was Sirdar of the Egyptian Army she visited Egypt, and it was her idea to found an Egyptian gallery in the Royal Institution, bullying her brother into providing a mummy as its nucleus.

Mary became ill in March 1894 and was rushed into hospital in London, where she died. Her body was brought back to Swansea for burial, and a large crowd followed her coffin on foot to Dan-y-Craig cemetery. The *Cambrian* put the numbers at around a thousand, which was probably an exaggeration. Even so, it must have been one of the largest funerals ever seen in 19th century Swansea.

THE MORRIS FAMILY

With the Baths, Vivians, and Grenfells, the Morris family was the fourth cornerstone of the Swansea copper industry in Victorian times. Morriston was the creation of the Morris family, and inseparably linked with the beginnings of the copper industry in the Swansea Valley.

Colonel Grant Francis gives the Llangyfelach Copper Works at Landore, owned by Dr. Lane and Mr. Pollard, as being in existence in 1717. In May 1727 the works came under the ownership of Robert Morris, in partnership with Richard Lockwood, E. Gibbon, and R. Barker. The works employed 40 men and two boys and the copper it produced was sent to London via Swansea and Bristol. There were no tramways in existence then: coal, ore, and copper had to be transported by carts or water – river or sea – or in baskets on the backs of horses or mules.[7]

By the mid 1840s Morris and Lockwood had established their new works at Fforest, which had the advantage of being nearer to the coal mines owned by Robert Morris at Trewyddfa. By 1764 Robert Morris had surrendered the lease of Llangyfelach, and when he died in 1768 his second son John assumed control and it was he who subsequently created the Morris industrial empire. He was born in July 1745, and married Henrietta, the daughter of Sir Phillip Musgrave of Cumberland. She was considered one of the twelve most beautiful women in the United Kingdom, and was painted by Romney. Around the year 1770, and by then a Baronet, John Morris built himself a Palladian-style mansion – Clasemont – at Llangyfelach, with panoramic views to the south-east.

Like the Baths, Vivians, and Grenfells after him, John Morris tackled the problem of housing his workers at subsidised rates. His first venture was in 1750 and consisted of an extraordinary castellated building on Graig Trewyddfa which provided accommodation for fifty families – the first flats for workers in Wales – perhaps in the world. It served its purpose for more than a hundred years, until local quarrying activities made it unsafe. Between 1790 and 1796 he went much further, creating what was perhaps the first of the model towns. It was designed by the celebrated bridge-builder of the day, William Edwards, and built near Nantfelin, not far from the Tawe River, and called Morristo(w)n. It was laid down in a rectangular grid pattern, each dwelling being provided with its own small garden.

He endowed St. Johns Church which stood centrally in the square, and is still there today. John Morris supported the 'Established Church' (no doubt for good business reasons) but he also assisted his Non-Conformist workers to build their own chapel. Average rents were 7/6d per year, based on ground leases which ran for fifty years (or 'three lives').

Common land on Trewyddfa was also made available to his most steady employees so that they could keep a cow, and he set up a form of Workmen's

7. Col. Grant Francis, *The Smelting of Copper in the Swansea District of South Wales from the time of Queen Elizabeth to the Present Day*, page 96.

Compensation Scheme which provided sick and injury pay, and payments in old age. Houses for the poor were built by him in both Morriston and Sketty. There was also a company shop where, it must be said, employees were expected to purchase most of their grocery and other household requirements, and this was associated with a system of tokens issued partly in lieu of wages. Today this produces unpleasant suspicions of worker exploitation by a Victorian employer, but again it has to be judged by the standards of the day. Without work people were reduced to the horrors of the workhouse; and starvation, illness, and death were always in the wings. People who had a home and permanent employment in Morriston considered themselves lucky. He died in 1819, and his son, Sir John Morris II, 1775-1855, who had lived in Sketty for many years, decided to build himself a house, Sketty Park, overlooking the Bryn area of Swansea. The 1823 *Swansea Guide* describes it as 'in a delightful situation, commanding the whole of Swansea Bay, and fine views over the country'. An important advantage, perhaps not trumpeted at the time, was its situation vis á vis the prevailing winds; like the Vivians, Sir John II no doubt preferred to avoid the noxious fumes belched out by his works, from which, philanthropic employer though he might be, he could not protect his workers. Sketty Park nevertheless did not have the imposing style of Clasemont. The house still stands, restored and renovated, being used nowadays largely for meetings and conferences, and housing an excellent restaurant.

Sir John II was the main instigator of the Mumbles railway, and its first chairman. Colliery owner, copper manufacturer, shipowner, town planner, railway pioneer, philanthropist and Town Councillor, he was another of Swansea's larger-than-life entrepreneurs. He was a humanitarian, and in many ways long before his time. The eldest son of Sir John II, Sir Armine Morris, was born at Bryn in 1813. Educated at Westminster and Sandhurst, he served in the Army for a time in the 60th Rifles, and subsequently became a Justice of the Peace. He lived for a while in France, letting out Sketty Park. Eventually he returned, and built himself a small house in its grounds, called Hafod, where he spent the last thirty years of his life, virtually as a recluse, devoting himself to Celtic Studies. Hafod was demolished in the 1890s and is now the site of a primary school. The home of his son, Sir Robert Armine Morris at Mumbles, Marine Villa, is now the British Legion club in Newton Road.

THE RICHARDSON FAMILY

The Richardson family are particularly interesting because their activities embraced the whole spectrum of the Swansea copper industry. They were shipowners, shipbuilders, repairers, brokers, timber importers, and copper handlers all on a scale which made them one of the leading merchant families of Swansea.

The founder of the dynasty, for so it must be termed, was John Richardson, born in North Shields on Tyneside in 1790, where he married (in South Shields) Elizabeth Clark of Leith (1791-1864) in 1809. He was connected with a prominent family in County Durham, Northumberland, where they had long been estab-

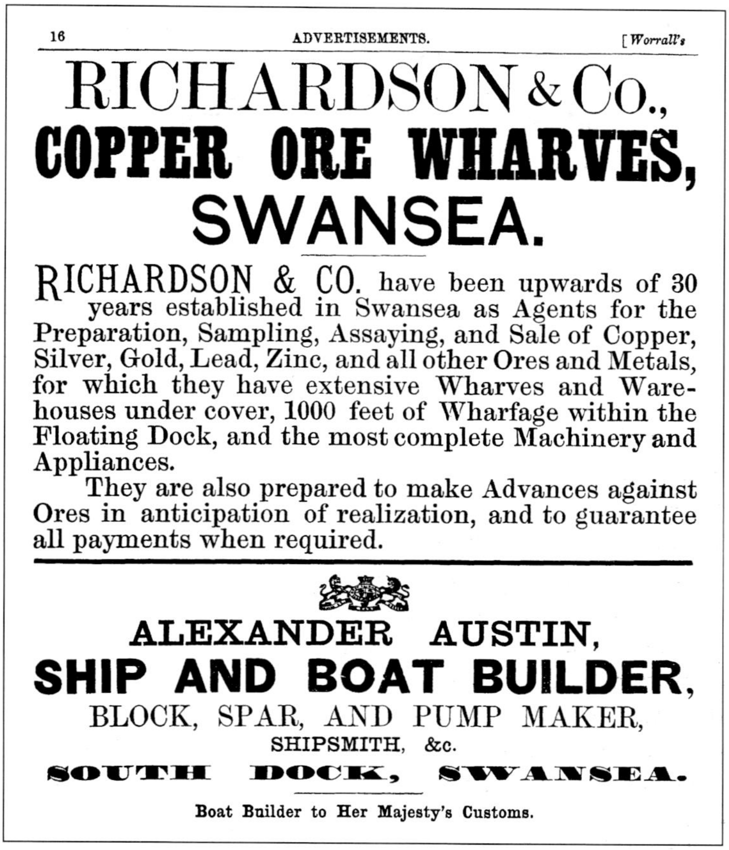

16 ADVERTISEMENTS. [Worrall's

RICHARDSON & Co.,

COPPER ORE WHARVES,

SWANSEA.

RICHARDSON & CO. have been upwards of 30 years established in Swansea as Agents for the Preparation, Sampling, Assaying, and Sale of Copper, Silver, Gold, Lead, Zinc, and all other Ores and Metals, for which they have extensive Wharves and Warehouses under cover, 1000 feet of Wharfage within the Floating Dock, and the most complete Machinery and Appliances.

They are also prepared to make Advances against Ores in anticipation of realization, and to guarantee all payments when required.

ALEXANDER AUSTIN,

SHIP AND BOAT BUILDER,

BLOCK, SPAR, AND PUMP MAKER,

SHIPSMITH, &c.

SOUTH DOCK, SWANSEA.

Boat Builder to Her Majesty's Customs.

Advertisements in Worral's 'Directory of South Wales and Newport', 1875.

lished as shipbuilders, and was a Master Mariner. He owned vessels engaged in the copper trade between Swansea and Cuba and in 1826 moved to Swansea in order to be able to manage his fleet on the spot.

He was involved in importing timber from the Baltic and Eastern Canada, and by the late 1820s had also commenced business as a shipbuilder, having obtained the lease of the old dry dock first established by the Meagers on the town side of the docks. Later he set up a timber yard along with a dry dock and 'Patent Slip' on the east side of the river Tawe, which subsequently became the Albion Dry Dock when the North Dock was built in 1852.

From their yard the Richardsons launched such well known vessels as the *Duke of Beaufort*, *Owen Glyndwr* and the *Marquis of Worcester*. All three were oak-timbered, copper-fastened, and A13 at Lloyds (i.e. considered by Lloyds to be a good risk for 13 years). All were put on the stocks in the late 1850s to provide employment during periods when repair work was slack, and took about three years to complete. They were intended for the copper trade to Cuba and Chile, and later traded to Port Nolloth in Namaqualand in Cape Colony. The *Marquis of Worcester* was followed by the *Lord Marmion*, 720 tons, the largest vessel then so far built in Swansea. At the same time the Richardsons were having larger vessels built to their order in Bideford. They also owned another well-known Swansea vessel, the *Marshall Pellissier*, 743 tons, and built in the USA. She was commanded in turn by Captain Thomas Ford and Captain Thomas Davies, both of whom later became Mayors of Swansea. Between 1870 and 1874 the Richardsons constructed two brigs of about 420 tons; these were the last they built; thereafter they confined their activities to shipowning and repairing, and copper factoring, which continued until about 1896, when they faded from the scene. In their hey-day the Richardsons were important copper factors, handling large quantities of ore for Pascoe Grenfell & Sons, the Vivians, William Foster, and the Cape Copper Company. The latter operated a large copper works known locally as Red Jacket, after an old public house near Briton Ferry.

The Richardson's fleet included such household names in Swansea, such as the *Glamorgan*, *Tacna* and *Corsair*, to name just a few. Their fleet list (see Appendix I) is subject to the almost universal caveat applying to Swansea-owned vessels, that of shared ownership, and the passing of ownership either down or among different members of the Richardson family. John Richardson's home, The Laurels, stood at the foot of Mount Pleasant, later the site of the Swansea fire station, demolished in 1997. During 1844-45 he was Mayor of Swansea and 1855 a Justice of the Peace. He died 21st November 1858, and was succeeded by his two eldest sons, John Crow (1810-1884) and James Coxon (1817-74), in managing the fleet and copper interests. His three younger sons, Jeremiah Clarke, Robert, and George managed the shipbuilding and dry dock side of the family business. John Crow Richardson lived at Pantygwydr House in the Uplands, but also had a seat at Glanbrydan, near Llandeilo.

Jeremiah Clarke, 1822-1906, followed his father as Lord Mayor of Swansea in 1864/5. During his mayoralty year he had to handle a particularly difficult situation when the copper barque *Hecla* arrived in the port with three of its crew dead from yellow fever. Fourteen days later yellow fever broke out and spread rapidly through the town, which in those days did not employ a medical officer of its own. The Mayor and his fellow magistrates ruled that stringent quarantine measures were not justified. It was put about that the *Hecla* was a Richardson owned vessel carrying copper ore to their yard, but this was shown to be untrue; they did not own the *Hecla* which was in fact chartered to the Cobre Mining Company. Nevertheless, it was a difficult time for Jeremiah Clarke, although he had personally initiated medical action and an emergency drainage scheme.

Jeremiah Clarke was a notable local yachtsmen, owning two well-known yachts, the *Kelpy* and the *Condor*. In 1904 he was elected the Commodore, of the Bristol Channel Yacht Club. His eldest son, John Crow II, became High Sheriff of Glamorgan in 1893.

THE BURGESS AND SHADDICK FAMILIES

The founder of the firm was James Edward Burgess, who began life as an apprentice to an iron-monger, John Beynon. He left without completing his apprenticeship in order to work for the Richardsons in the accounts department of their shipbuilding business. In 1843, when he was 21, he married Mary Francis Baker, whose father was a Barnstable miller and grain merchant who owned a bakery and provision business in College Street, Swansea. With the opening of the North Dock, the number of vessels using the port dramatically increased, and ship's biscuits, though a seemingly humble commodity, represented big business, since they were a staple part of the Victorian seaman's diet. Recognising the potential, James Edward invested in a mechanical bakery, the first of its kind in South Wales, which he set up at the bottom of Welcome Street in the Strand.[8]

By 1851 he was deeply involved in importing timber from the Maritime provinces of Canada, and in 1862 purchased his first vessel, the *Sarah Fox*, intended for the copper ore trade between Swansea and Chile. In the same year he also placed an order with W. Westacott at Barnstaple to build another copper barque, the *Patagonia*. In 1867, James Edward formed a partnership consisting of himself, his son, James Henry, and George Shaddick, who was later to become his son-in-law. A good idea of the scale of the new firm's activities is given by W. H. Jones in his *History of the Port of Swansea*:

> "Burgess and Shaddick . . . extended their interests over the whole field of transport and dock work, so that the name of Burgess quickly became of note in the shipping circles of south Wales. The firm may well claim to have been among the early promoters of the general cargo trade from Swansea to the United States by direct steamers in about 1879; they inaugurated direct trading between Swansea, Singapore, and China, and have often been the pioneers in opening up new ports for tinplates and other local productions. Still later, they added to their activities the exportation of coal, and today the firm of Burgess & Co Ltd. occupies one of the most prominent positions amongst the business firms of the port."

Many of the Burgess vessels were built at Barnstaple, and brought to Swansea for final fitting out and, most importantly, coppering. Between 1867 and 1877 the company bought or managed 23 vessels. In 1872 the *Standard Bearer* arrived in Queenstown from Iquique having completed the voyage in 70 days. The outward passage to Valparaiso took 71 days, so that the round voyage was completed in seven months, one of the fastest on record.

8. N. L. Thomas, *Swansea Districts and Villages*, page 140.

In the early 1900s George Shaddick, as Managing Director of Burgess & Co. Ltd., Steamship Owners and Agents, was Consul in Swansea for Portugal and Venezuela. The firm then had regular lines of steamships serving France, the Black Sea, St. Petersburg, Portugal, USA, India, China and Japan.

THE MEAGERS AND JOHN AND GEORGE RICHARDS

The Meager and Richards families played a key part in the development of shipbuilding and shipowning in Swansea. William Meager senior was a Cornishman, the son of Richard Meager of Bodmin, and was born in 1768. He and Edmund Richards were both working at the Devonport Dockyard in about 1812 when they decided to take their skills to the United States to commence shipbuilding there. Family tradition has it that while en route to the United States the vessel on which they were both passengers experienced severe storm damage and put into Milford Haven for repairs. In the meantime conversation with fellow passengers had changed their minds about emigrating to the U.S., and instead, hearing that William Jones's dry dock was to let in Swansea, went there to settle and set up their business. Their names first appear in the harbour accounts in 1812 in connection with paying for 'a mooring buoy, copper-fastened, charge £23. 5s. 2d.'

William Meager had two sons, George (1795-1850) and William (1800-1863) who joined him in the business. When their lease of the dry dock expired, the Meagers and the Richards then built the Phoenix Dry Dock, which extended into the river, and transferred their building and repair business there. Work at the Phoenix Dock continued until they had to relinquish the site when the Great Western Railway was built along the Strand. The partnership between the Meagers and the Richards appears to have been disolved in 1840.

In June of that year William Meager launched a new schooner of '93 tons register' for Mr. C. H. Smith of Gwernllwynwith. It is related of William Meager that while the old dry dock was in operation, repairs had been carried out on a foreign registered vessel, the master of which was suspected of planning to sail his vessel out of the dock at high tide, opening as it did, directly on to the river before settling his account. William Meager effectively forestalled this plan by taking the vessel's cables through the yard 'under the Oystermouth tramway lines and into the rear premises of the Pelican Inn, where they were chained and locked to a large baulk of timber placed across the back doorway'. Needless to say, his bill was paid before the vessel sailed.

In 1818 George Meager married Mary Hopkins and they had ten children. George Blaney Meager (1822-1877) took over the family business and after the opening of the North Dock in 1852 built the Villiers Dry Dock on land leased from the Earl of Jersey. His brother David Meager (1835-1920) was active in the ship-repairing business. He joined Swansea Council in 1865 and also founded and taught at a ragged school for poor children. In 1914 he became a J.P. By then the Meagers had become substantial shipowners. Like many other local shipowners, the Meagers were concerned in businesses directly connected with supplying the needs of sailing ships. In Pearse's *Swansea Directory*, 1856, Meager Bros., Strand, are listed as ship chandlers.

Writing in 1922, W. H. Jones mentions the schooner *Pitho*, 156 tons, built by William Meager and launched in 1868, which was still floating and working in 1919. Among the many well-known vessels built at the Villiers Dry Dock were the brigantine *Eliza*, 220 tons, for James Strick, the *Clara Manning*, also 220 tons, and the famous brig *Stranger*, 287 tons, launched in September 1859, specially built for the Cape Horn trade. Other well-known Cape Horners were the barque *Powhattan*, 399 tons, and the barque *Hamlet*, whose figurehead, carved by the Swansea woodcarver Levinson, was said to be an excellent likeness of Mr. George Melville, a noted actor of the day at the Star Theatre in Swansea. Apparently, in the best traditions of theatrical endeavour, Melville posed for the sculpture. Other vessels were the *Crown Thorpe*, *Maxima* and *Taunton*, but these by no means complete the list. The Meagers maintained their tradition of new enterprise into the 80s and 90s when they bought ships to run on the nitrate and guano trades to Chile and Peru.

THE MADGE FAMILY

The name Madge first appears in 1715 when the Corporation of Swansea granted a lease of a bank 'the breadth of his house on the Strands from the causeway to low water to a Robert Madge'. Later, in 1718, the same land was leased to William Gillar, an anchor-smith, to erect a forge or dwelling house, or both, before Michaelmas 1719, but he was 'not to prejudice the River or obstruct the way to or from the ferryboat'.

In 1831 another Madge appears in the person of a Captain Madge of the *Webster*, 129 tons, as one of a group of interested parties advising the Port Trustees on the establishment of a 'New Cut' to convert both the river and harbour into 'independent but connected floats of water'.

During the 60s Henry James Madge constantly had vessels building at Bideford in North Devon, and his barque *Capricorn* was one of the first copper ore vessels to be built there in 1859. Another well-known Swansea barque, the *Serena* (see page 144) belonged to Henry Madge, as well as a brig named the *Bedford* which had a square poop and stern windows.

Perhaps the best known of the Madge vessels was the *Chelydra*, 339 tons, built at Newport in 1850. She had been constructed with the Chinese opium trade in mind, and was fitted with guns, presumably with piracy in mind, and comfortable passenger accommodation. The story goes that on entering Swansea she would fire her guns to announce her arrival.

WILLIAM JENKINS ('BUTCHER' JENKINS)

William Jenkins was a prominent Swansea shipowner in the 1840s. His money was derived from the business he conducted in St. Mary Street as a cattle dealer and butcher. Most of his vessels were built in Bideford, and subsequently brought to Swansea to be coppered. His vessels included the barque *Alicia* which was lost on her first voyage, and her name subsequently given to another of his vessels. He owned another vessel, the *Rajah of Sarawak*, 525 tons, built by George Cox at Bideford in 1850 and A13 at Lloyds. She was specially constructed for the East India trade, and her 'tween decks could be fitted up to carry troops. She was considered to be so much in advance of her time that people travelled from as far afield as Merthyr Tydfil just to view her. Other vessels owned by Jenkins were *Catherine Jenkins*, *Kate Kerney* and *Cuba*.

There is a story that one of Jenkins' vessels never returned from her voyage, and it was suspected, but never proved, that her Master had renamed and re-painted her, and either sold her or used her on his own account.

His eldest son, also William Jenkins, joined him in the business, and sadly died aged thirty in May 1851. Another son, George, was drowned, aged twenty, on 12th February 1843 off Mauritius, while travelling home from China on the *Matilda* of Liverpool. William Jenkins himself died 7th February 1850 aged seventy, and was buried in St. Mary's churchyard in Swansea, close to the eastern gate.

ISAAC GALE

Isaac Gale was in business in a big way as family grocers and provision merchant at 18, High Street. There was extensive storage for provisions, wine, and spirits, and also a bottling department, together with facilities for blending tea and coffee. The firm specialised in fine brands of coffee, which was roasted in London and sent down to Swansea where it was blended. By the mid 1800s the business was actually owned and conducted by B. Goodhall, though still operating under the trade name of I. Gale. Naturally, a major part of the firm's business consisted in supplying stores to ships, and like many other local businesses with similar connections, the firm owned part shares in many of the vessels which they supplied.

I. Gale was also well-known for supplying top quality china and porcelain – Dresden, Majolica, Worcester, and Japanese ware, which was displayed in a spectacular showroom on the second floor of their establishment.

BERNARD RUDKIN HENNESSEY

Bernard Rudkin Hennessey was a well-known Swansea watch, clock, and chronometer-maker who was in business in Wind Street between 1842 and '85. Born in

Dublin, he was also a Gun-maker (in 1858 he was the sole agent for 'Colonel Colt's celebrated pistols'), Optician, Supplier of Scientific Instruments, Jeweller and Contractor to the Admiralty. He was also a supreme self-publicist; the pages of the *Cambrian* newspaper are filled year after year with reports of presentation cups, plates, salvers, and medals, mostly made or designed by himself. He was the archetypal Victorian dynamo who combined practical industry with commercial skill.

In March 1864 he presented a 'valuable 8-day clock and thermometer' to the Swansea Sailor's Home. Many of his activities were already closely connected with shipping, especially the adjustment of ship's chronometers and sextants.

He is another good example of a local businessman whose activities led him into shipowning almost as a matter of course. In 1877 he is listed as one of twenty-three local businessmen owning 2,000 tons or more of shipping in the port. Among these perhaps the best known were the Swansea barques *Antarctic*, *Arctic* and *Tacna*. In the Swansea *Evening Post*, November 1956, Mr. Dick Sullivan, one of the then few surviving Swansea 'Cape Horners', said in an article, 'I stowed away in the *Tacna*, a barque belonging to Mr. Hennessey, the jeweller of Wind Street. Unfortunately before we passed Lundy I was discovered and, broken hearted, I returned to Swansea in the tug-boat.'

HENRY HOSKINS

Was a shipowner and potato merchant whose business was situated at Padleys Yard and who resided at Uplands House, Uplands. He owned a considerable number of vessels, many of which were built in Prince Edward Island and engaged on the nitrate and copper trade to Chile and Peru. Among these were the *Bride*, *Bridegroom*, *Bridesmaid*, *Cwmdonkin* and *Harry*. Hoskins always named his vessels after his family and its locality.

SIMON GOLDBERG & CO

Simon Goldberg was a pawnbroker who had his business at 17 Goat Street, Swansea. He invested heavily in shipping and came to own one of Swansea's most important fleets, which included such famous local vessels as the *Agnes Lillian*, *Vanduara* and *Brynymor*.

The *Brynymor* is particularly interesting because she comes at the end of the period covered by this book, and was one of the largest vessels to sail out of Swansea.

The *Brynymor* was a full rigged ship of 2131 tons, built in 1902 at Dumbarton by Macmillan and Company, and was largely engaged in the nitrate trade to Chile and Peru. For her first three years, until she was sold in Hamburg in 1905, she was commanded by the redoubtable Captain James Bevan (see page 126), then Commodore of the Goldberg fleet, who had sailed round the Horn fifty-six times.

JOHN AND WILLIAM BOWEN

The Bowen brothers, John and William operated from Swansea and Llanelly respectively. John Bowen owned, among others, the *Ida*, *Raven* and *Morna*, all of about 200 tons, and built in Prince Edward island in the 1870s. He also owned the two-masted brig *Ottawa*, 225 tons. William Bowen in Llanelly owned the ex-tea clipper *Lothair* and the *Jessie*, *Kate Helena* and *Georgina*.

SIMPSON BROTHERS

Captain J. H. Simpson was sent to Swansea by the Sunderland shipowners William Ord and Co. who had a large fleet of barques running from Swansea in the Cuba and later Chile copper ore trade to represent their interests in the port. These included such vessels as the *Hilda*, *Ianthe*, *Charles Lambert*, *Stately*, *Florence* and *Adelaide*. He later went into ownership on his own account with such vessels as the *Pembroke Castle*, *Picton Castle* and *Benchruachan*.

THE SWANSEA MERCHANT SHIPOWNERS COMPANY

This company was set up in 1873 with a nominal capital of £250,000 in £50 shares with the object of taking over the fleet owned by Henry Bath & Co., the main investors being Jeremiah Clarke Richardson and Edward Bath. Other important investors were Frank Ash Yeo and John Glasbrook, both colliery owners. The shares were widely spread around the Swansea district – farmers, drapers, hotel-keepers, fishmongers, and private gentlemen. James Harris Jnr., the Swansea marine painter and Cape Horner, was among them.

The company reduced its capital in 1888 as the copper trade went into decline, and it was wound up in 1895.

*

There were many other shipowners in Swansea, about whom little can be said within the scope of this book, other than to list their names and the ships they owned, and these, together with those already mentioned appear in Appendices I and II.

5.

Swansea Port in the 19th Century

COMING INTO SWANSEA FROM CARDIFF and the east, Kilvey Hill rises 600 feet to form a backdrop to the town. From its summit the surrounding landscape lies stretched before the viewer. At its foot are Port Tennant and St. Thomas and, on the other side of the Tawe river, Hafod. In the 19th century almost every other house in all three parishes housed a seaman's family. To the west, where the harbour stretches out into the estuary, is Swansea Bay, once described by William Savage Landore as 'better than the Bay of Naples' at high tide, with Mumbles Point and its lighthouse at the far end.

To the north-east is the valley of the Tawe, where in Victorian times 90% of the world's copper was smelted, then wreathed permanently in a pall of smoke and the acrid smells of heavy industry.

On the other side lies the estuary of the River Neath, then with its own foundries and heavy engineering works, and beyond that again the tree-clad valleys of Aberdulais and Glynneath. The three great centres of gravity around which 19th century industry mushroomed in Victorian times were St. Thomas, Llansamlet (the Grenfells), Hafod (the Vivians) and Morriston (the Morris family).

In 1832 the town boundaries previously established by parliament were redefined in order to incorporate within the borough the suburbs which had grown up around each centre of industry. The total population was then about 19,000 and by 1851 it had risen to nearly 32,000. St. Thomas, Morriston, Pentre-Clwyth, Landore and Manselton still retained their character as separate communities, while the town of Swansea itself, along the river and beside its castle, was an intricate complex of shipowning, shipbuilding and repairing, engineering, chemical and manufacturing, and business activity.

As early as 1819 the town was connected by regular steam packet and sail traders to Bristol, Gloucester and London. At Gloucester goods could be shipped onwards by canal to Worcester, Birmingham and the industrial areas in the north. The comprehensive nature of the network is graphically illustrated by the following extracts from Matthew's *Swansea Directory* of 1830:

Canal Packet Boats,

Leave PORT TENNANT, every evening from Four to Six o'clock, for NEATH; a reasonable and convenient conveyance for Passengers and Goods.

STEAM PACKETS.

THE BRISTOL STEAM PACKET,
JOHN EDWARDS, Commander,

Of 60-horse power, leaves Swansea for Bristol, with goods and passengers, twice every week, and returns from Bristol, and the following days, and to and from Ilfracomb, once a week.

Agents, Mr. C. Marten, 4 Fisher street, or Packet Office, Ferry house, Swansea; Messrs. Rees and Williams, 12 Quay street, or to Mr. James Huxtable, 22 on the Back, Bristol; or Amos Fry, Packet house, Ilfracomb.

THE PALMERSTON STEAM PACKET,
JOHN HYDE, Commander,

Of 86 Horse-power, leaves Swansea for Bristol, with goods and passengers, twice every week, and returns on the following days, and to and from Ilfracombe once a week.

Agents, Mr. S. Padley, Strand, Swansea; Messrs. Terrell and Sons, Back, Bristol and Steam Packet Company's Office, No. 1, Quay, Bristol.

COASTING VESSELS.

LONDON TRADERS.

Brothers, T. Lewis | *Henry*, William Thomas
Eliza, James Thomas | *Sarah*, John Thomas.
To and from Hay's Wharf, Tooley street, London.
Agent, Mr. Joseph Lewis, Somerset place.

Paul, John Webster | *Jane*, William Davies.
To and from Cotton's Wharf, London.
Agent, Mr. R. Walters, Wassail street.

BRISTOL TRADERS.

Eleanor, Thomas George—*Rose*, John Jenkins
Phœnix, William Lodge—*Swansea Trader*, P. Huxtable
Swansea Packet, William Barrett.
One of these Traders are to sail every fourth day, taking their regular turns in loading.
Agent, Mr. Joseph Lewis, Somerset place.

GLOUCESTER TRADERS.

Takes goods for Worcester, Stourport, Birmingham, and all parts of the North.
Abeona, W. Jones—*Halcyon*, D. Tanner
Sarah, W. Pocket.

Agents
J. BROMLEY, Worcester Wharf, Birmingham.
J. C. BROADFIELD .... Stourport.
H. SOUTHAN .... Worcestor.
E. WALKER .... Gloucester.

Office, at the *Tiger*, Strand.

PACKET.

The *Packet*, ~~Amos Fry~~, Master, sails every Tuesday and Friday, for Ilfracomb.

***Statement** of the number of Vessels, with their Cargoes, which have entered Inward, and cleared Outward, at the port of Swansea, in the years* 1812 *and* 1819, *respectively.*

CARGOES AND NUMBER OF SHIPS.

FOREIGN, INWARD.

	Years. 1812.	1819.
Number of ships	36	96
Loads of timber	229	888
Deals and deal ends	85	41
Loads of masts and spars	.	7
Number of ditto	.	204
Fathoms of lathwood	25	48
Cwt. of copper ore	4,660	50,929
Heads of cattle	1,297	549
Pigs	1,362	70
Horses	.	30
Cwt. of potatoes	3,200	1,285
Yards of linen	16,250	58,375
Packages of general goods	65	303
Quarters of corn	2,865	3,271

FOREIGN, OUTWARD.

	1812.	1819.
Number of ships	417	449
Chalders of coals	15,943	15,434
Chalders of culm	10,021	11,844
Tons of stone coal	379	167
Pieces of earthenware	50,993	140,2[illegible]
Tons of Copper	10	101

FOREIGN, OUTWARD.

	Years. 1812.	1819.
Tons of bark	128	230
Packages of general goods	103	169

COASTWISE, INWARD.

	1812.	1819.
Number of ships	943	1,556
Slates	488,601	664,242
Tons of stone	310	1,178
Gallons of wine	4,577	6,864
Gals. of Foreign & British spirits	13,532	15,566
Pounds of tobacco	45,458	34,420
Pounds of snuff	8,094	12,112
Bushels of salt	3,994	4,474
Tons of ditto	37	28
Chests of tea and coffee	830	812
Hogsheads of sugar	475	484
Tierces of ditto	434	441
Lumps of refined sugar	1,485	1,284
Packages of grocery	917	1,187
Chests of candles	208	307
Chests of soap	435	681
Tons of hemp	62	85
Quarters of malt	4,520	4,044
Quarters of corn	6,685	7,469
Quarters of flower	2,737	5,092
Tons of lead	24	27
Boxes of tin	91	190

On the edges of the borough were the collieries, quarries and brickworks which provided the raw material for its growth. In the town itself were the shipowners, copper-works owners, bankers, lawyers, doctors and merchants who were responsible for propelling Swansea into its position as the copper-smelting centre of the world.

The social structure was complex. At the top were the landed gentry, headed by the Duke of Beaufort. Although the largest owner he did not live within the area and controlled his interests through agents. The layer below consisted of what might be described as the gentrified industrialists – powerful local families whose money and influence had propelled them upwards in society, and occasionally into the nobility.

In between these two layers was an urban gentry of whom perhaps the best example might appear in the person of Colonel Grant-Francis, F.S.A., J.P. His title of colonel derived from the commission he held as colonel of the Glamorgan Artillery Volunteers, and he was a distinct local character of a type which the Victorian age seemed to specialise in producing. His home was at Cae Bailey at the foot of Mount Pleasant Hill, and his name constantly appears in local reports relating to anything that might concern the improvement of the town and its port. He was largely instrumental in setting up the Royal Institution of South Wales, founded in 1835 as the Swansea Philosophical and Literary Society, and became its president. Among the other founding members were Lewis Weston Dillwyn, John Henry Vivian and C. R. M. Talbot, all Fellows of the Royal Society. He was Secretary of the Dock Company, and was, among many other things, largely responsible for persuading the War Office to install a battery of guns at Mumbles at a cost of £10,000 'to defend the roadstead'. His book, *The History of Copper Smelting in the Swansea District*, published in 1881, was an authoritative contemporary account, and his last great contribution was to push through the railway line which linked Swansea to the Rhondda Valley in 1882.

The majority of Swansea's massive workforce were immigrants to the area, or their children. At the bottom of the social heap were the Irish, largely condemned by prejudice to the worst jobs and poorest housing. But in the outer borough and especially in the upper valleys, the workers were predominantly Welsh and Welsh-speaking. The communities were all working-class in character, housing, habits and dress, and for most of them life was extremely hard. Their work was often dangerous, and nearly always injurious to health. The whole working-class population tended to be thin, sallow, and suffering from lung diseases of one sort or another, aggravated by a poor diet, and what we now know as stress, a condition scarcely recognised by doctors in those days. Their recreation, depending on the outlook of the family concerned, consisted largely of the practice of non-conformist religion or drinking, as the number of chapels and public houses eloquently testified.

By 1875, the population had exploded to over 150,000 and Swansea had become one of the leading sailing ship ports in the world. In 1849 the Swansea Board of

Health had commissioned a report on conditions in the town, and a few extracts speak volumes: '. . . the worst parts are probably about the northern end of High Street and Greenhill Street, and their subordinate courts, especially the district known as Little Ireland. In Green Row the privies are in a filthy condition'. Mr. Davis, who compiled the report, goes on to say: 'I have been informed that as many as 30 have lodged the same night in one of these cottages. They are a nuisance to the neighbourhood.'

A crowded South Dock, Swansea, in Victorian times.
(Courtesy Swansea Museum)

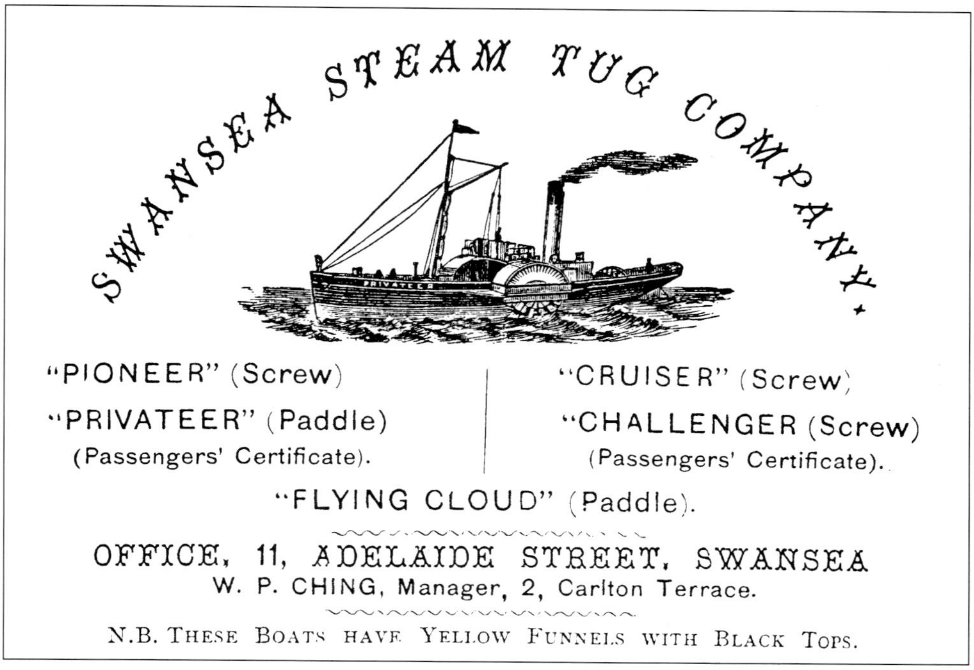

Business card of the Swansea Steam Tug Company.
(Courtesy John Tanner, Swansea)

Copper barques in South Dock, Swansea, circa 1870s.
(Private Collection)

In Mariner Street, he reports that there were forty-three houses with 30 privies. 'The floor of one is so inundated by the liquid contents of the pit, so that the seat can only be reached by means of large stones.' The water supplies were appalling, mostly from stand-taps which were cut off for five or six hours almost daily.

There were two well-known springs, one behind the Royal Oak and the other in the Strand, both largely choked with debris and rubbish. What sewers existed were frequently clogged and overflowing, and the system discharged into the river, where it was only inadequately carried away by the tide. Only the main streets of the town were reasonably surfaced, with flagged pavements.

Swansea, like most major seaports in the world contained its seamy side, much to the horror of its more respectable citizens. The *Cambrian*, 9th February 1877, carried a lengthy editorial denouncing the existence of 'Disorderly Houses' in thundering terms larded with such phrases as 'the blameless purity of home life which is the Palladium of our continued greatness . . . on behalf of the high minded burgesses of Swansea who are the majesty of the law . . . The recent raid, or vigorous effort to put down brothels, and to suppress disorderliness in the town is not commenced a moment too soon . . . moreover the denizens of these iniquitous houses walk the streets unabashed in broad-glare . . . Chief among such places is Wind Street where SHAME parades herself undisguisedly, in the latest hours of the night'. The utterances of such high minded middle class persons are often inadvertently comic, even today; one hopes that charity was also not forgotten.

Before 1852, when the Tawe river was diverted by the 'New Cut' to form the North Dock, vessels had for centuries discharged or loaded their cargoes at the wharves which lined the west bank, lying as best they could on the muddy, tidal estuary. The area was known as the Strand, and it was lined with public houses, warehouses and small businesses.

There were alleyways which wound their way up to Wind Street and High Street, which lay above it, and these housed a rabbit warren of centres or 'courts' where people lived. There were dealers in marine stores, rope and canvas makers, hairdressers, grocers, basket workshops, tailors and boot-makers. Later on, when the railway line arrived in 1850, its terminus, as now, was at High Street Station. Another line was built in 1852 along the Strand, supported by brick and stonework, and punctuated by arches. From that period onwards the Strand became a road; the river no longer flowed past it after being diverted by the New Cut and the arches housed a multiplicity of specialised trades connected with the sea. There was Levinson, the famous carver of ship's figureheads, anchor-smiths, ship-chandlers, wheel-wrights, boiler-makers, ship-coopers, block-makers, brass-founders and paint suppliers, to name only some. The whole area formed a sort of seething mass of ceaseless activity.

Pearse and Brown's *Swansea Directory* of 1869 gives an interesting list of tradespeople directly concerned with supplying ships requirements:

Block and Spar Makers
T. Evans, 35 Langdon Place
Mrs. W. Evans, 13 Adelaide Street
T. Robinson, South Prospect, Strand
G. Taylor, 7 Upper Trafalgar Terrace
J. V. Thomas, Strand
Joseph Webb, Pier Street

Chain Cable Maker and Anchor Smith
George Davies, Strand

Rope Maker
D. Williams, 43 Waterloo Street

Sailmakers and Dealers
Johnson and Son, Broad Quay
Charles McAdam, Bath Lane
Charles McNamara, Adelaide Street
Evan Phillips, 30 Bellvue Street
William Phillips, Castle Cottage, Cambrian Place
W. Pike, Lion Stores, Quay

Ship Biscuit Makers
T. Trow, Crown Factory, Strand
Tucker, Smith & Co., Srand

Ship Chandlers
George Ace, Ferryside
Johnson & Son, Broad Quay
David Jones, Strand
H. P. Langlois, 4 Prospect Place
James Madge, Broad Quay
V. Morrison & Co., 4 Fisher Street
L. & G. Tulloch, Strand and Gloucester Place

In some places, the Strand formed a kind of rear extension to the large businesses in Castle Street and Castle Square, like John Taylor the 'high class provision merchant' whose premises had a frontage of thirty-six and a depth of two hundred and fifty feet.

Into this maelstrom of commercial activity came the crews of the copper barques, heading for the twenty-five or more public houses on the Strand or in its vicinity. The captains frequented the respectable houses like the 'Mackworth Hotel',

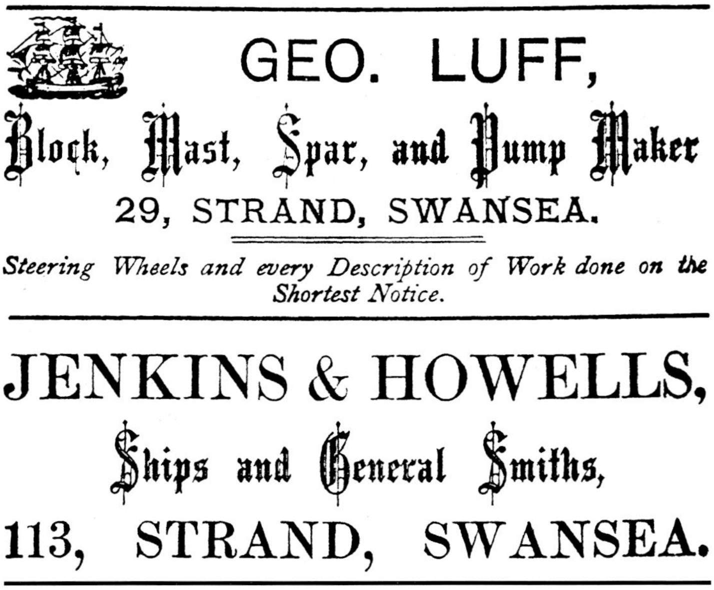

GEO. LUFF,

Block, Mast, Spar, and Pump Maker

29, STRAND, SWANSEA.

Steering Wheels and every Description of Work done on the Shortest Notice.

JENKINS & HOWELLS,

Ships and General Smiths,

113, STRAND, SWANSEA.

Ex Butchers Swansea Directory, 1875-6.

'Castle Hotel' or the 'George Hotel'. There was a sort of unwritten law connected with the hierarchy on board ship, which by common consent was generally observed. For instance ship's officers, as distinct from Captains, tended to patronise the 'Cameron' which stood on the site now occupied by Argos.

It is tempting to list all the public houses frequented by the crews but such a catalogue, and of names only, would be too long. By common consent, the public bar of the 'Mackworth Hotel' was open to the crews. The most celebrated seamen's pubs were 'Dirty Dicks', 'Troubadour', 'Kings Arms', and the 'Ship Hotel'. Others which were also boarding houses, included the 'Cardiff Arms', which, as these words are written in 1999, still stands, forlorn and shuttered, and with its paint peeling off but otherwise intact. Other well-known hostelries included the 'Liverpool Arms', 'Pelican Inn', 'Welcome to Town' and the 'Green Dragon'. Their very names are redolent with the vanished world of the sailing ship, and it is not difficult to imagine the sort of atmosphere they must have possessed, thronged with seamen who had sailed to the far corners of the world.

Padley's Yard, leading off the Strand between numbers 129 and 130, deserves recording. It was Swansea's Covent Garden, where potatoes, barley meal, salt and every kind of vegetable was sold. There were even two Breton onion-men living in the yard, and it contained the offices of Henry Hoskins, Potato Merchant, who was one of Swansea's leading shipowners (see page 63). Beau Nash was born in Goat Street, and when George Borrow walked through Salubrious Passage, (still there today), he remarked that it was distinctly 'insalubrious'.

A common site in the town on Saturdays and Wednesdays were the cockle women from Penclawydd, dressed in Welsh costume and sitting on small stools, selling cold cockles, shelled or unshelled.

The most active periods for shipbuilding in Swansea were the late 1850s, 60s and the early 70s. In the 60s the Meagers were building three ships at a time, while in the early 70s there was the Richardson's Wharf on the New Cut side. Further on was their dry dock where more than one vessel at a time could be built or repaired. Until the end of the 19th century, Swansea port would have had a characteristic quality quite different from twenty years later. As Basil Lubbock wrote in his *Last of the Wind-Jammers*, throughout the 1860s and 70s the North Dock was 'jammed so full of little barques that you could walk across it in any direction by stepping from vessel to vessel. And on the other side (of South America) it was no uncommon thing to find 90 to 100 sail of Swansea copper ore men lying in the Bay of Callao at the same time.'

In February 1999, I had the privilege of interviewing Captain C. E. Mock, of Kittle a retired Swansea pilot, who though 102 years of age perfectly retains all his faculties, and especially his memory, although his vision is somewhat impaired. He began life as apprentice to Edwards, the draper in High Street, Swansea, but soon decided to go to sea, which he did as a mess-room boy. His time was spent in steam, where he obtained his Board of Trade certificates including Extra Master in the shortest possible time.

Captain Mock had a fund of information about life in Swansea port around the turn of the century, including such nuggets as 'Jerry' the one-legged rigger, who could be seen working aloft on the top-gallant yard, the empty leg of his trousers flapping in the wind, bending sails before hauling down to the jiboom corner. According to Captain Mock, the 'Green Dragon' or 'Shipping Gazette' were famous as the houses where thirsty crews returning after the long voyage from Cape Horn slaked their thirst. 'And what sing-songs they had at the 'Gloucester', all troubles forgotten for the time being.'

All the smells and sounds of a busy seaport would be intermingled; the rich smell of newly-sawn timber, the pervasive odour of tar, oil from the tanning of sails, horse-droppings from hundreds of carts, the stink of harbour mud at low tide, and the distinctive scent of salt water-soaked hemp rope. Against it all would have been the clinking of mallets on caulking irons, and the squeak from blocks and tackle being hauled on masts and yards.

Only the screech of seagulls and the river itself remain, together with the crumbling remains of a few warehouses against the brooding backdrop of Kilvey Hill and the ramparts of Swansea Castle. Supermarkets and car parks now stand where sailing vessels once floated and the men who sailed in them are long ago at rest in quiet churchyards, or in the sea itself.

Over the years the *Cambrian* newspaper is full of accounts of gales and storms in the Bristol Channel, and shipwrecks and strandings were common. Sometimes there were upwards of a hundred sailing vessels anchored in the roads off Mumbles,

The barque 'Valparaiso'.

The barque 'Andaman'.

and a sudden change of wind direction could spell disaster. One such occasion was reported in the *Cambrian* on 27th November 1824 as follows:

> "During the last three weeks we have experienced a continuation of rainy and boisterous weather, which drove a great number of vessels bound down channel, from Swansea, Neath, and other places, to seek shelter at the Mumbles, where on Monday last, upwards of 100 were observed riding at anchor in perfect safety from the then prevailing weatherly wind. On Monday night, the wind suddenly veered to the south-east, blowing almost a hurricane, with heavy rain and thunder, when a scene of the most indescribable confusion and dismay ensued at that place among the shipping in the darkness of the night. In the morning, it was ascertained that the damage, though considerable, was not of the extent which might have been expected from so sudden a tempest, and we are happy to add, that there were no lives lost; fourteen vessels were observed to be on shore, more or less damaged, but with the exception of the *Lavinia Johns*, of Falmouth, and a Bideford sloop, have been got off."
>
> The following is a list of vessels which have come to our knowledge, as having sustained damage: *Fame*, Curtis, of Hayle, with loss of boat, cable, and bulwarks; the *Amelin*, Millwood, of Hayle, with loss of bowsprit and bulwarks, and stanchions carried away; *Elizah*, Williams of St. Ives, with loss of bowsprit, rudder, gib, and main boom, and bulwarks; *Redruth*, Row, of St. Ives, with loss of bowsprit; vine, *Shelton*, from Newport, with loss of bowsprit, &c.; *Union Courts*, of Fowey, with loss of three anchors, cables and bulwarks; *Emerald*, Walters, of Fowey, and *Rebecca*, Stibley, of Padstow, with sterns stove in, and damaged; *Rasleigh*, of Melhuish, with loss of three cables, gib-boom, and bulwarks; *Uterphe*, Watley, with windlass bitts carried away, and copper bottom injured; *Mary Ann*, of Padstow, with loss of bowsprit and anchors; *Howe*, Care, of Penzance, with loss of bowsprit; and the *Sophia*, Melhuish, with loss of bulwarks and boat. – The *Catherine Roberts*, of Pwllheli, and the *Assiduns*, Trick, of Swansea, received trifling damage. – A Dutch galliot arrived in our port, on the above evening, considerably damaged, and in great distress.
>
> In a severe gale, on Saturday se'nninght, the sloop *Plain Dealings*, John Jones, master, was lost on the Lavan Sands, near Penmaenmawr, and the crew, consisting of the Captain, one man, and a boy, perished.
>
> A large ship, supposed to be a West Indiaman, has been wrecked in the neighbourhood of Mount's Bay, a quantity of rum and coffee having been picked up about the Bay, washed on shore by the sea.

On Thursday morning part of the hull of a vessel drove on shore under Marazion cliff, with masts, booms, cabin doors, a great deal of light timber, and many other things. She is supposed to have had a general cargo, but where she struck or belongs to, cannot now be ascertained, none of the crew having been saved."

The 'Illimani' off the Mumbles, by Alexander Brander.
(Private Collection)

The barque 'Epsilon' in Swansea, May 1888.
(Private Collection)

6.

The Shipbuilding Triangle: Swansea, North Devon and Prince Edward Island, Canada

THROUGHOUT THE BOOM YEARS of the Swansea copper trade, some vessels were built in Swansea, notably by the Richardsons with such well-known craft as the *Duke of Beaufort*, *Owen Glyndwr* and *Marquis of Worcester*, or in Sunderland, Glasgow, Liverpool, Cardiff and Southampton and other places. But most were built either in North Devon, or Prince Edward Island in the Maritime Provinces of Eastern Canada.

The forests of Canada and North America provided virtually unlimited supplies of wood, albeit mostly softwood, and on Prince Edward Island (P.E.I.) in particular there was a strong cultural connection with North Devon and Swansea.

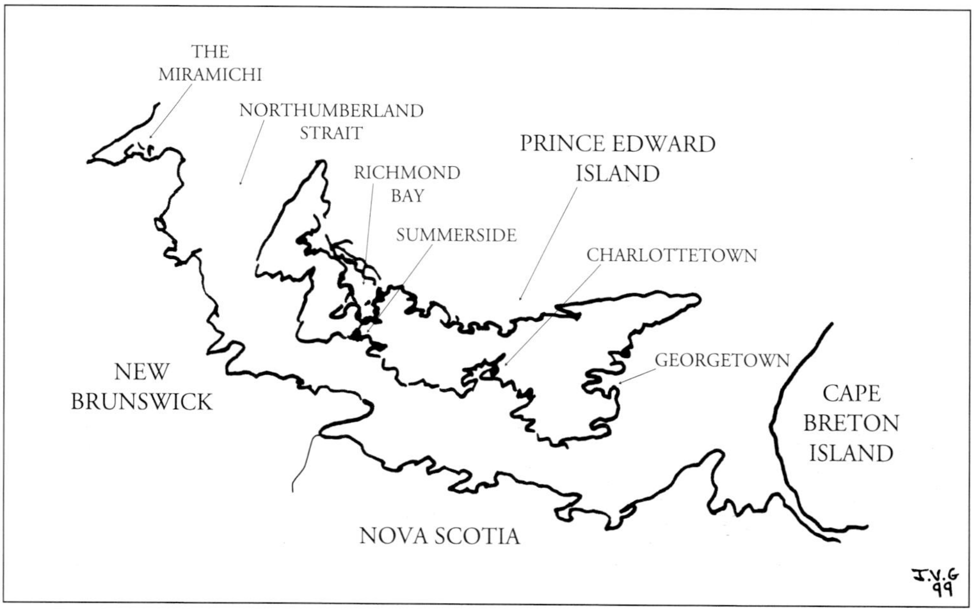

Sketch map of Prince Edward Island, Canada.

Bideford, with its associated port of Appledore, basically on a creek of Bideford, together with Barnstaple, to the north-east, at the end of its own creek, were natural centres for wooden shipbuilding. Swansea, almost directly opposite across the Bristol Channel (people then living in west Gower often found it easier to cross the channel to Appledore or Ilfracombe than journey to Swansea by horse and cart) required vessels in ever increasing numbers to meet the growing demands of its copper-smelting industry. 'Coal out and copper ore in', shipowning, smelting and ship-repairing had turned Swansea into one of the leading sailing ship ports in the world.

These three areas, drawn together by ties of culture and business interest, interacted and cross-fertilised each other, and shipbuilding in one cannot be described without bringing in the other two.

That stretch of the North Devon coast which lies between Lynmouth Bay and Hartland Point is considered by many to be one of the finest stretches of coastline in the British Isles. It is not dissimilar to Gower, being a rock-bound coast, indented by coves, that could spell disaster for a sailing vessel driven on to it by north-westerly gales, and behind, a hinterland of hilly rolling moors given over to farming and walkers.

The little seaport town of Appledore stands at the head of a peninsula between the river Torridge and the Bristol Channel. Where the Torridge flows into the sea from the south, it is joined from the north by the river Taw, and both flow into the sea through a gap in the Bideford Bar, a huge sandbank that lies across the river mouth between the Northam Burrows to the north, and an outlying bank of sand to the south, known as The Neck.

Even today, with powered craft, passage of the Bar requires careful navigation; to a sailing vessel it was always treacherous, and in other than good conditions of wind and tide, exceedingly dangerous.

Where the waters of the Torridge and Taw meet and mingle, there is a deep tidal pool where vessels could float at high tide and lie comfortably on the mud when it was out. It was even crossed by a cart road, which wound its way among the mooring posts. Around this pool grew up the town of Appledore, situated on a creek of Bideford. In the 1750s, Thomas Benson of Northam, a businessman who had extensive trading connections with North America, built the first stone quay on one side of this pool, two hundred feet long, and thus spawned the town of Appledore as it is today. Still with its narrow streets and alleys, pastel washed houses and quayside, it is not difficult today to visualise the little seaport as it must have been in Victorian times; its streets and quay crowded with townsfolk, all connected with the sea; shipbuilders, sail-makers, rope makers, ship-chandlers, shipowners and shopkeepers. Shipbuilding in this area first began in the 16th century. Quoting from the National Maritime Museum's publication *Shipbuilding in North Devon* (1976): 'Leland, visiting about 1540, noted, in his highly individual style, that at Bideford there was "a Praty Quik Streat of Smithes and other occupiers for Ship

Crafte".' Then in 1566 a ship of 250 tons was built at Bideford for an Exeter merchant. Finally, in the last decade we find the names of two specific vessels built at Barnstaple in 1593. These, among accounts of payments of the Royal Bounty of 5s per ton, were the *Galleon Duddeley*, 400 tons, and the *Roger Bonaventure*, 120 tons.

A brigantine under sail.
(Private Collection)

As in Victorian society generally, Appledore was very much divided up into social layers, with the local landed gentry perched at the top. In Appledore itself the urban gentry and larger shipowners and merchants lived at the top of the hill; the lesser ones and businessmen half way up, while at the bottom, in a rabbit warren of little houses, alleys and courts above the beach, dwelt the seamen, shipwrights, sawyers and boatmen.

In 1810 Moses Chanter bought Staddon, a fine Georgian house at the top of the hill, previously occupied by a shipbuilder, Richard Chapman. He married Elizabeth Burnard, by whom he had a son, Thomas Burnard Chanter, who was destined to play a significant role in developing North Devon's connection with P.E.I.

Moses Chanter was a substantial merchant and shipowner, and Bideford like other shipbuilding ports in the United Kingdom was suffering the effects of the blockade against Britain imposed on continental ports by Napoleon in the Berlin Decree of November 1806. The Baltic ports had been a major source of timber and supplies in July 1807 suddenly ceased exporting to Britain following the Treaty of Tilsit, when Russia, Prussia, and Denmark joined the blockade. Canada, Nova Scotia, New Brunswick and P.E.I. were all then under British rule; timber, including oak, was available from Vermont and New York via the St. Lawrence River, but the merchants in Quebec who handled it wanted a guarantee that when peace returned their investments would not be lost. The answer therefore lay in Canada, and especially in P.E.I., thickly covered with dense pine forests. Thomas Burnard, a prominent local businessman, dealt in timber and in October 1812 a vessel, the *Four Friends*, under the command of Captain Ramsey, arrived at Bideford with a cargo of softwood from P.E.I.

P.E.I. was first used by Basque fishermen and subsequently colonised by French settlers who were dispossessed by the British during the French wars. In 1796 the British government was bombarded with requests for land grants as new colonies were acquired overseas, and with a typical mixture of laxity and feathering the nests of its cronies, saddled P.E.I. with sixty-seven Army and Naval officers and supporters, who immediately became the absentee landlords of the island with all the potential for greedy exploitation that brought with it.

They acquired land at nominal rates – a few shillings per hundred acres, which they promptly farmed out to hard working settlers. The first governor, Walter Patterson, arrived in 1769, building his own log cabin in what is now the capital, Charlottetown. The only resources the first settlers had were farming, and the forests which covered the island. The basics of colonial life, such as ironmongery, fishing gear, cloth, seeds, sugar, oil, salt, and soap had all to be imported, and the merchants who handled these came to virtually control the island.

Charlottetown was a true colonial settlement, its few houses being little more than log huts, each surrounded by its own patch of grass where the owner's pigs and cattle could graze. The settler would grow his own spring wheat and barley, and kitchen garden produce like potatoes and other vegetables. If he was lucky he might own a pair of draught oxen. Their lives were dictated by the seasons; between January and March the timber would be felled and dragged over the frozen ground to the wharves from which it could be shipped.

William Ellis, born at Monkleigh in Devon in 1774, was a master-shipwright, and was sent out to Charlottetown in a small sloop-rigged vessel named the *Peter and Sarah* by Thomas Burnard with seven shipwrights and seamen to set up a shipbuilding yard there. They arrived at Charlottetown on the 13th September 1818 with enough ironwork, rope, sails, and fittings to build a 200 ton vessel.

There was one site on Richmond Bay which seemed suitable. Richard Moys, a shipbuilder, who may have been one of the party, obtained a 999-year lease of 313 acres bordering the Goodwood river from Sir James Montgomery. Existing settlers were paid to fell and transport enough timber to build the schooner and provide a cargo of lumber to take back to Bideford, where it could be used or sold.

William Ellis' first vessel, the *Mars*, was completed in 1819 and Richard Moys sailed in her as master to deliver the cargo at Bideford.

In 1824 Thomas Burnard died, having from poor beginnings built up a large timber shipbuilding and shipowning business. He was also a banker, and four times mayor of Bideford. More than anyone else, he was responsible for the founding of shipbuilding in P.E.I.

In 1820 the *Mars* had been sent out with enough material to build three more vessels, under the overall control of Thomas Chanter, Thomas Burnard's nephew. Thomas Chanter had come to manage Thomas Burnard's business interests in P.E.I., but when the latter died he left his property to his wife and eldest son, Nathaniel. Thomas Chanter was not mentioned in the will.

In 1824 Thomas Chanter arrived at Richmond Bay, P.E.I., and shortly afterwards it was announced that the Burnard shipbuilding business in P.E.I. would be known as Burnard and Company. The notice, which was signed by Nathaniel Burnard and Thomas Chanter, was issued from a house called Port Hill, named after a similar house built in North Devon by Augustus Saltren Willetts on a peninsula overlooking the river at Bideford.

The business was then sublet to William Ellis in 1825 and subsequently made over to Thomas Chanter, who entered into a contract with Ellis, on terms very favourable to himself, whereby the latter ran the shipbuilding business and also occupied Port Hill.

Eventually, in 1845, William Ellis retired. By then he had become a sub-tenant of James Yeo. William Ellis's business ability unfortunately did not match his skill at shipbuilding. He died on Christmas Day 1855, widely respected throughout P.E.I., aged 82.

At this point there enters upon the stage the formidable figure of James Yeo of Bideford, who came to dominate shipbuilding in P.E.I., and more or less everything else as well. He was born at Kilkhampton, near Bideford, the son of James Yeo senior, a local shoe-mender. It is known that around 1815 he got together enough

James Yeo, Senior.
(Portrait by Robert Harris, Museum and Heritage Foundation, Charlottetown, P.E.I., Canada)

money to acquire a horse and van, and set up a weekly carrier service between Kilkhampton and Bideford. If we take his age as then being about twenty he would have been born circa 1795.

In 1818 his wife Anne died; James began drinking and the carrier business failed. But it had brought him into contact with Thomas Burnard, who must have detected potential in James, for he paid off his debts and dispatched him to P.E.I. to act as his agent in gathering together cargoes of timber for shipment to Bideford.

James was accompanied by a second wife, Damaris Sargent, who turned out to be just as shrewd in business as he was. For it soon became obvious that James, although he could only barely read and write, possessed an excellent business sense, and moreover was a natural genius at figures and accounting, aided by a phenomenal memory.

One thing is certain, he prospered in P.E.I. and in due course managed to get William Ellis into his debt, and ultimately acquired his shipyard. By the 1840s he was the island's most prosperous shipbuilder and merchant, and owned a fleet of locally built vessels, some from his own yard, and others at the different yards which by then had sprung up on the island.

His son William Yeo by his first wife, sailed for several years in command of his father's vessels, though it remains something of a mystery how he actually obtained a master's certificate. The links between Bideford in Devon and P.E.I. continued to be strong, and eventually William Yeo was sent to Bideford to represent his father's interests in Britain. As soon as William was established there the old man began to make regular visits to Devon, and even London.

William Yeo was no less ruthless than his father and achieved a somewhat similar reputation in Bideford. If it was not quite so extreme that was probably only because the opportunities for exploitation in Bideford were not as great as they were in P.E.I.

The vessels built at P.E.I. were mostly sent to Devon (or later Swansea) for finishing off, e.g. re-caulking, additional copper fastening on the hull, and copper sheathing on the bottom.

By 1855, the Yeos had acquired sufficient land in the valley which ran down to the Quay in Appledore to build a large dry dock, together with associated shipbuilding facilities, including a foundry, smelting, mould-loft and sail-loft. Finally, on the hill above the dry dock, they built an imposing residence, a potent symbol of the power and success of James Yeo, and the arrival of the Yeos as one of the leading merchant families of North Devon. It was complete with terraced gardens, extensive grounds, and all the paraphernalia of squiredom – lodges coach-houses, and a high stone wall. On 17th July 1856 the barque *Elizabeth Yeo* entered the dry dock to the firing of guns the shouts of a large crowd.

It was said of James that in 1838 he was the only man in P.E.I. from whom the settlers could obtain ready money (against a mortgage). Forty years after his death, people there still shuddered at his name. He was ruthless, physically strong, and hardly ever slept – he was said to live more or less on horse-back, supervising his enterprises. He seemed to care little for his children; it is said that he did not even know their birthdays. But his tribal loyalty in terms of business was strong.

Damaris Yeo died in 1867 of pneumonia, and James a few months later from the same cause.

William Yeo died in 1872 at 59; his death created a depression in Appledore which lasted for ten years. In P.E.I. the four wheeler James Yeo had used in his later years was left in the barn, the whip still in position. Damaris's clothes still hung in her closet. They were still there when the house was demolished after the First World War.

Bideford and Barnstaple had a reputation for high quality shipbuilding. 10 year ships (vessels regarded by Lloyds as insurable for ten years) were standard by the 1830s, and increased to 13 years by the 1850s.

The chief shipbuilders at Barnstaple were the Westacott family, whose shipbuilding began with William Westacott in 1796. The earliest vessel known to have been built by them was the brigantine *Elizabeth*, 68 tons, built in 1814 by John and Robert Westacott. Robert either died or left the business around 1820 and John became the master builder until his death in 1850. Their first barque was the

Roderick Dhu, 359 tons, built in 1845 for the Swansea copper ore trade and thereafter they built barques almost continuously for Swansea until 1871. In that year they launched their last barque, the *Standard Bearer*, 480 tons.

As the demand for wooden sailing vessels fell, the yard became more and more dependent on ship-repairing. Their last vessel, other than small craft, was the brigantine *Rosina*, 194 tons, in 1884, ordered by a Bristol merchant, Charles Bennett, who sold her in Newfoundland.

Also at Bideford was the Johnson's Yard, taken over by Robert Johnson around 1840. He died in 1855 aged 61, and his wife carried on the business until his son John took it over in 1858, under the title of Robert Johnson and Son. Their first barque was the *Hugo*, 369 tons, built in 1862 for London owners, and this was followed by several well-known barques for Swansea owners. The best known of these was the *Florence Danvers*, 498 tons, built 1865 and the *Beatrice*, 455 tons, in the same year.

The third well-known Bideford builder was George Cox, born in Bridport (1811-77) and his son John (1834-79). The Coxes built barques and larger vessels such as the *Emperor of China*, 485 tons, in 1847 and *Rajah of Sarawak*, 525 tons, in 1850. Between 1858 and 1866 they built more than a dozen barques for the Swansea copper trade.

Their yard at Cleave Houses had been taken over around 1825 by Thomas Evans, who later took George Cox (1811-77) into partnership about 1841. Evans seems to have relinquished his interests around 1845, leaving George Cox and his son John in partnership. When George Cox died in 1877 the yard closed down and John died two years later.

The sheer numbers of vessels built in North Devon are impressive. Between 1850 and 1880 a total of 377 were built.

The high quality of North Devon shipbuilding may have been equalled elsewhere, but it was never bettered. The longevity of the wooden vessels they produced were a bye word. During both World Wars North Devon made a major contribution in shipbuilding, and the Swansea connection with Appledore for wooden boats continued into the 1950s. Swansea owners bought eighteen foot 'National' hulls from Hinks, and in 1958 P. Waters & Sons built the ketch *Dorothy and Mary* for Swansea owners.

North Devon auxiliary trading ketches and schooners regularly visited Swansea North Dock basin until 1960, including such well-known ketches as the *Garlandstone*, *Emily Barratt* and the schooner *Kathleen and May*.

Today the Mumbles Yacht Club's tender *Kenda* built by Hinks at Appledore incorporates and, as it were, preserves, in Swnsea the handed down knowledge of generations of North Devon wooden shipbuilders.

The chief pioneer of the connection between Swansea, Devon and P.E.I. was William Richards, Master Mariner, born in Swansea in 1819. He went to sea at the age of eighteen and by 1844 was in command of the 175 ton brig *John Hawkes*, built on P.E.I. She was partly owned by Joshua Williams of P.E.I., who later sold her to the Swansea shipowners, Beynon and Tucker.

The Tuckers had an agent in P.E.I. in the person of J. M. Tucker. William Richards became acquainted with James Yeo in 1845 when the *John Hawkes* loaded timber at Richmond Bay, P.E.I. and in 1849 he married Yeo's second daughter, Susannah. In 1851 he was employed by James Yeo in command of various of his vessels. In 1853, William's brother, Thomas Picton Richards, by then a Swansea shipowner, came to P.E.I. to buy a brigantine which he named the *Queen* (which in December 1867 was lost in horrific circumstances on Gull Island, Newfoundland.

Thomas Picton led an adventurous life, ranging from being in California during the gold rush in 1849, and walking across the Isthmus at Panama at the time the railway was being constructed. He became one of Swansea's most important shipowners, although in his early days he sometimes acted as a super-cargo on his own vessels sailing to the Mediterranean. In later years he went into partnership with Samuel Browning Power under the trading name of Richard Power and Company. One of their most famous vessels was the *Darwin*, a 1,000 ton barque, which carried a figurehead carved by Levinson, the celebrated Swansea carver, which was said to be a very good likeness of the famous author of the *Origin of Species*. He died at an advanced age in 1915, having handed the business over to his son, T. P. Rose-Richards, who was High Sheriff of Brecknockshire in 1921/2.

In 1858 William built his first vessel, the *Richmond*, 284 tons, at Grand River, which was transferred to Swansea, like so many of the vessels he subsequently built. His first six vessels, apart from the *Richmond*, which he financed himself were jointly financed with his father-in-law. James Yeo. After that he conducted business on his own account and between 1862 and 1892 built 34 vessels, all near his home in Bideford. Among these was the celebrated barque *Maggie*, 402 tons, in 1867, A6 at Lloyds of which there is a model in the Swansea maritime museum. His most productive years were 1874, '75 and '77, building three vessels in each year. William Richards had two sons, James William and John, who in due course joined him in the business.

Excluding the six vessels he built in co-operation with the Yeos, William Richards and his sons built in all: 3 schooners, 5 brigantines, 6 brigs, 11 barque and 14 barquentines. All were transferred to Swansea to his brother Thomas Picton Richards, who deployed them mostly in the copper ore trade. Thomas Picton, as well as being a shipowner, was very active in ship-broking. Of 246 vessels constructed on the western side of Richmond Bay, only 57 were built by other than Thomas Burnard, Thomas Bernard Chanter, the Yeos, and the Richards, and nearly all of them were transferred to the UK.

While other areas of shipbuilding in P.E.I. actively suffered from the vagaries of non-resident investment, the Yeos and the Richards made Richmond Bay their domain and the primary centre of shipbuilding in P.E.I.

The following tables are derived from research material contained in Nicolas de Jong and Marven E. Moore's classic work, *Shipbuilding on Prince Edward Island.*

MAIN SHIPBUILDERS AT P.E.I. 1860-80

Builder	*Place*	*Built*	*Transfer UK*	*Period*
Duncan & Co.	Charlottetown	89	65	1860/80
Hyndman & Co.	Charlettetown	21	9	1869/80
John Lefurgey	Summerside	18	12	1869/80
Longworth Bros.	Charlottetown	65	50	1860/80
MacDonald	Georgetown	17	9	1869/80
Angus McMillan	Summerside & Egmont Bay	51	41	1860/80
Owen and Welsh	Charlottetown	133	95	1860/80
Peake Bros.	Charlottetown	87	64	1860/80
W. Richards	Bodeford	74	61	1860/80
Richards and Yeos together	Port Hill	116	105	1860/80

BREAKDOWN OF VESSELS BUILT AT P.E.I. 1869-80

Year	*Schooner*	*Brigantine*	*Brig*	*Barquentine*	*Barque*
1869	11	22	11	–	11
1870	12	25	9	1	10
1871	16	22	7	–	9
1872	24	22	10	–	7
1873	10	31	8	2	13
1874	13	35	14	4	22
1875	14	33	12	10	11
1876	11	25	6	9	7
1877	13	21	11	3	11
1878	9	10	5	5	6
1879	6	9	–	–	4
1880	11	6	–	3	1
Total	150	261	93	37	112

In P.E.I., spruce trees now thickly grow along the water's edge where the last ship was launched before World War I, and the ruins of Chanter's Wharf still juts out from the old farm at Port Hill. P.E.I. is now a popular tourist attraction for town-weary folk from the rest of Canada, and its white houses are beautiful against the lush green countryside.

I cannot leave this chapter without recording how much it owes to Basil Greenhill's definitive work, *West Countrymen in Prince Edward Island.*

7.

The Building of the Barques

SWANSEA SHIPOWNERS were frequently shipbuilders, and vice versa, and it is not easy to neatly separate them and still provide a coherent description of both. And so often the ownership of a vessel was divided up into shares, involving a cross-section of businessmen, farmers, colliery owners, local landowners and gentry. The larger owners like the Baths and Richardsons not only owned their own vessels, but also bought shares in others. Ownership was shared between different families, individuals, and enterprises. Nominal ownership of vessels was passed down or between members of the same family, or in and out of collective ownership. Local businessmen would build or buy a vessel, or part of it, and then sell it, sometimes only to buy it back again later.

The vessels which traded regularly out of Swansea were also owned in London, Liverpool, Sunderland, Devon, Somerset, Prince Edward Island in Canada, and other places, even as far away as Chile. They come and go, and reappear, swirling about, so to speak, in the tides and cross-currents of Victorian entrepreneurship. The major advantage of divided ownership was the sharing of insurance risk, for the vessels were engaged in the most dangerous trade in the world, and the cost of casualties, losses, and repairs was high.

Of particular interest, as described in Chapter 6, were the close links which existed between Swansea, Prince Edward Island, North Devon and Liverpool, where so many of the Swansea-bound copper ore vessels returning from Chile had first to discharge part of their cargo because of the limited draught offered by Swansea port prior to 1852.

The stage is a crowded one; endeavouring to describe the actors in alphabetical order would create a kind of dictionary approach, neither desirable or explanatory of their place in the time pattern, and equally, their relative importance. So they will appear as a combination of both.

Early builders in Swansea were Henry Squire, who leased 'banks or yards for the building of ships situated on the Strand' from the corporation in 1765; William Jones, who built a dry dock in 1776, and Isaac Griffiths, who erected a shipbuilding yard on the Burrows near the Ferry House in 1780.

In the early 1800s George Llewellyn built a number of vessels in his yard at

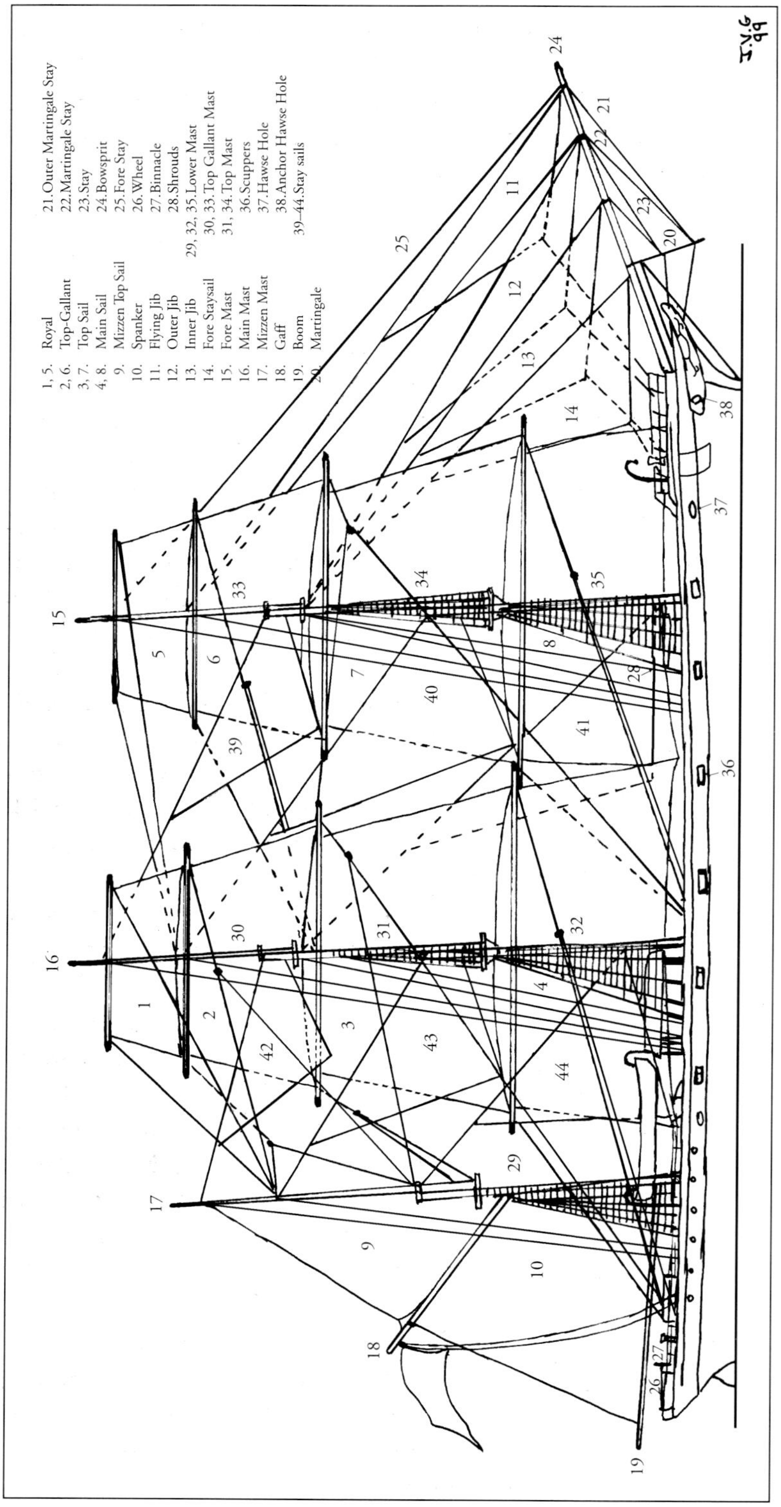

Fig. 1: Sketch plan of a typical Barque.

Swansea, the best known of which was the *Cambrian*, 200 tons. While commanded by Captain William Cook, who was also her chief owner, the *Cambrian* encountered the East Indiaman *Kent* on 1st March 1825, which had been set on fire because a cask of spirits had been set alight by a fallen lamp. A total of 557 persons out of 642 on board were rescued and put ashore at Falmouth.

Around 1812 his yard was taken over by William Meager and the brothers John and George Richards, and with them shipbuilding in Swansea can be said to have begun in earnest.

The copper barques, along with the brigs, barquentines, topsail schooners and full-rigged ships, large and small, which in Victorian times dotted Swansea Bay, are so interesting today because they represent so eloquently a vanished world of which we are still immensely aware. It was a world where a small local community could build beautiful ships and sail them to the far corners of the world.

Small vessels like smacks and schooners could and often were built all round the coast of the United Kingdom wherever there was space to build a vessel and enough water to launch it. Little yards like these could be started up with very little capital, all of which would be forthcoming locally with many shareholders joining in the risk. The materials required were largely available locally, and the main ingredient, labour, was on the spot and prepared to work for very low wages.

The skills required would all be local; basically, carpenters, blacksmiths, sail-makers, and rope-makers. They would live locally, probably in cottages with no running water or toilet facilities other than a thunder-box at the bottom of the garden. The luckier ones might have enough ground for a pig or few chickens, but they all worked long hours for little return. Yet they had immense pride in their work, and were respected members of the community.

In such small enterprises there would typically be a slipway, a forge, a sawpit, and a shed to store materials. In larger enterprises such as those situated in seaport towns like Swansea, Bideford, Cardiff, Bristol, and dozens of others including Prince Edward Island in Eastern Canada, there would be, in addition to a forge and sawpits, a mould loft and sail-maker's loft, and a considerable store of timber, either ready for use or left to season. Some of this, like oak, elm, and beech, would be available in the surrounding area, but increasingly, as timber supplies in the UK became scarcer, supplies were imported from the Baltic, North America and Canada. There would also probably be a local rope-walk, and other specialist trades, such as figurehead carvers, anchor-smiths, chain-makers, mast and spar-makers, blockmakers, tree-nail makers, and riggers.

About twenty to fifty men might be employed, and there are many photographs of such groups, often taken on launching day, with the Owner and Master Builder clearly distinguishable by their attire from the yard foreman and head moulder wearing bowler hats, to the youngest employee in short-sleeves with folded arms. They are similar to the photographs of the ship's officers and crews who sailed the ships they built, which confront us today in fading sepia mounted on thick yellow

The bow of the 'Patagonia' at Swansea, circa 1880.
(Private Collection)

The bowsprit and figurehead of the 'Olivebank'.

cardboard with the photographer's name in flowing italics on the back. They stare out at us, symbolising the Victorian age with all its confidence, vigour, enterprise; proud, some defiant as if in spite of their obvious poverty, visible reminders of a mercantile manufacturing world now vanished forever. Yet its image remains, somehow comforting and satisfying to remember as surrounded by televisions, computers and portable telephones, we contemplate our own problem-ridden world.

At the start of the 19th century the typical merchant vessel had changed relatively little over the preceding two centuries. Built of wood and rigged with natural fibres, the hull shapes were always rounded, so that they could sit upright on the mud in tidal harbours. Their blunt bows and general lines made them sail very slowly, and hardly at all to windward. But bustling Victorian entrepreneurship, boiling down as it did to cargo-carrying capacity, running costs, and speed, soon led to improvements in ship's lines.

Special impetus was given to this by the fruit trade, where oranges and lemons were imported in huge quantities from ports all round the Mediterranean. They were mostly schooners or brigs; 250 tons was about the largest, but the lessons learned from their design, mostly empirically, were applied to larger vessels like barques and ships. Another important factor was the influence of the American builders, who were far ahead of the British in designing for speed. Cross-fertilisation, especially from the Maritime Provinces of Canada, a prolific source of economically-built vessels, played an important part.

The 'Fruiters', as they were called, really deserved the name of clipper, and this may be an appropriate point at which to refer to that frequently contentious designation. The definition I use throughout this book is that of Harold Underhill, who felt that a clipper was 'any reasonably fine-lined vessel that was fast in relation to the average craft of her time, a craft which given similar conditions could outsail the then everyday merchantmen . . . so the clipper of one age could be the slow-coach of the next.'

'Tonnage' is another term which can usefully be clarified at this stage, since it crops up with every vessel. Originally it derived from the cost of hiring a vessel at so much per ton (tun) of the cargo she could carry; otherwise known as its 'burthen' or 'burden'. This arose from a tax introduced by Edward III in 1347 on the import of wine at the rate of three shillings for every tun. In 1694 a law was passed in Britain requiring merchant vessels to carry a waterline mark which indicated the fully laden and empty of cargo levels (in ballast). This gave rise to a formula calculation which became known as 'Builder's Old Measurement' (BOM). In 1773 an Act of Parliament defined this more specifically with the formula $\frac{(L-\frac{3}{5}B) \times B \times \frac{1}{2}B}{94}$ where B is breadth and L = length.

However, another calculation became necessary since in order to arrive at what cargo a vessel could carry the space occupied by the crew's quarters, galley, stores, etc., had to be deducted from the gross tonnage in order to arrive at the net tonnage. The registered tonnage, which was entered on the vessel's certificate of registration, provided the basis on which harbour duties, towage charges, salvage assessments, etc., were levied.

Later, another measurement was introduced, that of dead weight tonnage, which was the amount of cargo a vessel could carry when loaded down to her plimsoll line mark (see page 119).

The formulae used for calculating these tonnages inhibited ship design, where cargo capacity was always a prime consideration, but in 1854 the restrictions imposed by the earlier ways of measurement were largely removed, and ship design rapidly improved.

HALF MODELS

In the days of wooden shipbuilding, at least until towards the end of the era, vessels were not planned out on paper, other than by rough drawings. Instead, a half model of the hull would be carved from a block of wood, which reflected the ideas of the builder, illuminated by his own experience, and advice from sea captains who knew how a vessel behaved in a seaway. The owner commissioning the vessel might be an ex sea-captain himself; if not he would probably have the opinions of his own masters to guide him. And all would be individualists with strong ideas on the sort of vessel they required.

A sailing vessel was a very sensitive creation – almost, one might say, a work of art, and it was difficult to predict what its performance would be. There were many cases of vessels being built specially for speed turning out to be slow, and occasionally one built with no great expectations in mind would turn out to be a flier.

On the origin of the half model itself there is an American claim, not accepted everywhere, that the idea of making a hull model which could be taken apart to provide builder's lines was that of Orlando Merril, a young shipbuilder of Newbury Port, U.S.A. in 1794. The original model made by him was presented to the New York Historical Society in 1853. He died in 1855 at the age of ninety, and it is said that he made no money from his invention.

The block would be laminated – yellow pine or mahogany was popular – in layers about ½" to ¾" thick, and its profile would represent the finished lines of the vessel to the outside of her frames, three to four feet being a common length. The planking was then added and secured, so that when the laminations were separated each plank provided the contour of the vessel at that particular point. The half model was of course to scale, so that each lamination could be used as a basis for measurement in the moulding loft.

Some of the half-models would be reassembled and kept for reference. Sometimes they were polished and mounted on a suitable base, with stumps of masts, bulwarks, and deck houses added. But many were thrown away or used as firewood. Today they are rare and much sought after as collectors' pieces.

THE MOULDING LOFT

The chief feature of the moulding loft was its floor, on which would be chalked out either a full or half-breadth plan of the vessel, taken from the model, on a scale dictated by the space available in the loft, The moulder, an all important character in the building of a wooden vessel, would start off with a straight line representing the length of the vessel. From this would be drawn other lines at right angles – twelve was a common number – spaced more or less evenly amidships but nearer together fore and aft where the curve of the vessel was more pronounced. The length of these lines, again taken to scale from the half model, would indicate the breadth plan outline at the twelve different points.

The moulder's next step would be to bend two flexible battens so that they touched these points from amidships to fore and aft respectively. The natural curve taken up by the batten was quite remarkable in the extent to which it produced a line very near to that which would be arrived at by calculation by a modern day designer. The dhows of the Persian Gulf and Red Sea have been built on the same principles for centuries, and lines taken off them suggest that their hull shape could scarcely be bettered for the purposes for which they are intended.

The next stage would be for the half-model laminations to be dismantled, one layer at a time, and drawn out to scale on the mould floor; the sum of this information would enable the moulder to then decide on the exact shape of each frame. Next full size two-dimensional templates were made up of each frame from scrap timber, the outer edge of which indicated the outer edge of the frames. These were numbered and thus when all in place reproduced the lines of the half-model on the finished vessel.

The importance of the moulder's job can scarcely be exaggerated; on him depended the ultimate success of the vessel and consequently that of the yard. His other vital responsibility would be to select the timbers used in the construction, and a key factor in this would be to take advantage of natural bends in the wood to form the innumerable curved parts of the hull. Natural bends conferred great strength at vital points and were an important element in wooden ship construction.

The timber in the yard together would be stacked and allowed to season; oak, elm, and beech from the UK, and pitch-pine, Oregon pine, and yellow pine from the Baltic. The planks were cut by sawyers working in a sawpit; only late in Victorian times did steam-driven saws become available.

SAWPITS AND SAWYERS

The sawpits were about eight feet deep and lined with logs or stone slabs. If the pit was near the water's edge, as was often the case, water seeped in, and the pit boy, who worked the bottom of the saw would sometimes work all day with his feet in water, perhaps in freezing weather.

The top sawyer provided the power stroke; the job of the pit boy was merely to guide the saw, while standing under a constant shower of sawdust. Eventually, after four or five years, he would become a top sawyer; hence the old saying, 'cocky as a top sawyer'. They were men of immense physical strength and endurance, and commanded much respect both for that and their skill. An experienced sawyer could cut through a long length of yard square timber to very fine limits, and all the timber used in the yard would be sawn by them.

Steam-powered saws did not come into general use until the 1870s, and then only in the larger yards. By the time they became general, wooden shipbuilding apart from small craft, was coming to an end.

The sawyers provided their own saws, which they had to keep sharp. Their working day was from six in the morning until six in the evening, with a break of about an hour at mid-day and four in the afternoon. In the winter, when light was bad at either end of the day they worked by naphtha flares. The timber was pushed along the pit top over rollers, and held in position by wedges. Hard work and hard times indeed; when unemployment or illness could often mean something very close to starvation.

THE SLIPWAY

This was where the vessel grew from first laying down the keel to the point where the hull could be launched. First the keel would be laid down on keel blocks on the floor of the slipway, which would be slightly inclined towards the water to aid the launching process. The blocks would be about four feet apart and about three feet deep, so that a man could work underneath them. Unseasoned elm was always preferred for the keel because it improved in saltwater and did not deteriorate as it dried out. The stem and the stern posts would then be scarfed on, massive deadwood being used to strengthen the whole structure.

Joining this mass of timber presented special problems. Iron or copper bolts were used, the latter being much superior though more expensive since it was more resistant to corrosion. To fit the bolts, holes were drilled by augers, some of which were as long as eight feet, and required turning for hours. The bolts were in bar, usually about an inch in diameter, and the hole was made slightly smaller and ensuring that it was true required great skill. The bolts were driven into the hole by gangs of men wielding sledge-hammers, a task again requiring skill, because the head of the bolt had always to be hit squarely. Both ends were then burred off, thus creating a gigantic rivet.

The frames were of massive construction, made up of perhaps six or more sections, with the bottom part always in one piece and sitting athwart the keel, with the curved portions being carefully chosen from naturally shaped timber. Finishing off was done with an adze, another highly skilled operation. Once again, bolts were used to go right through the frame and the keel, the head of the bolt being countersunk and packed round with oakum and putty.

At this point the keelson, an additional internal keel, would be scarfed together, care being taken to ensure that the joins did not coincide with those in the keel itself. It would then be butted on to the keel.

The massive nature of construction in wooden vessels was quite remarkable, Harold Underhill, in his classic book *Deepwater Sail* quotes an old seaman speaking of the barque *Shenandoah*: 'I remember standing in her limbers (spaces in the floor timbers of wooden vessels on either side of the keelson to allow a free passage for the bilge water to run down into the pump well) and finding the keelson coming up to my chin, and I am no pygmy, being five feet nine and a half inches tall.'

Allowing for the thickness of planking and depth of keel this would amount to more than six feet of solid timber in the keel and keelson. The next step would be to fit a line of timbers scarfed together, about ten inches thick and a foot deep, all round the inside of the hull, just below deck level. This was called the beam shelf, and on it was placed the deck beams, each supported by a wooden knee, and secured by copper bolts. Provision would be made within the deck beam structure for cargo hatches and hatchways giving access to the crew's accommodation, forecastle, cargo spaces, etc.

Finally came the planking up, the planks being first shaped and then steamed in a steam chest before fitting. Here again the skill of the yard moulder was vital; each plank would be tried out against the frames, with the fore end being shaped to fit into the rabbet and the inside of each plank adzed away on the inner side to fit the curve of the frame. When the planks were sufficiently flexible, they would be rushed into place by a planking gang and clamped in position ready for bolting. Several gangs would be at work and it was a matter of pride not to have to return the plank to the steam chest for a second steaming. The plank ends were always cut so that when they were butted the joint fell on a frame. The inner edges of the planks were shaped so that they touched each other on the frame and left a gap of about quarter of an inch on the outside for caulking. Such shaping called for many years of experience from a shipwright.

The deck planking was a great deal easier; the work was commenced amidships and developed from there fore and aft, and breadthways. The stanchions, the upright supports that carried the guard rail, would be scarfed to a frame head. These were always a weak spot in a wooden vessel; if they were carried away by a heavy sea water could pour into the hull, and especially if the vessel was deep laden, cause it to founder. Plugging the holes in the sort of weather that would carry the stanchions away was extremely difficult, situated as they were on the extreme edge of the vessel. The waterway, the thick piece of timber which ran all round the inside of the bulwarks at deck level, provided extra strengthening at such points, and was hollowed to provide a channel which would carry water of the decks via the scuppers. The scuppers were holes cut in the bulwarks to allow the deck water to run out; on the outside shutters hinged at the top prevented water from coming in. These were also known as freeing ports.

Finally, the caulkers would get to work with oakum and pitch. Each caulker had his own selection of specially shaped irons, and the sound of the mallet as it struck the iron produced a distinctive musical note. Caulkers would cut their mallets to produce their own note, and the chorus produced by the intermingling of these sounds was a characteristic of old Swansea seaport, now alas vanished forever.

Copper sheeting was essential on vessels sailing in tropical waters to their hulls from the ravages of the teredo worm and encrustation by weeds and barnacles. These could grow rapidly in the course of a few months – weeds several feet long which trailed out and made the vessel slow and difficult to steer. Barnacles had the same effect, and there could literally be many tons of them.

On iron-fastened vessels galvanic action between the copper plating and the heads of bolts created problems which could result in whole plates dropping off, hence the popularity of the more expensive copper fastening. The hull would be heavily tarred below the water line, and felt sheets, soaked with tar, pressed on to the hull contour. The coppers sheets were then fastened on with copper nails.

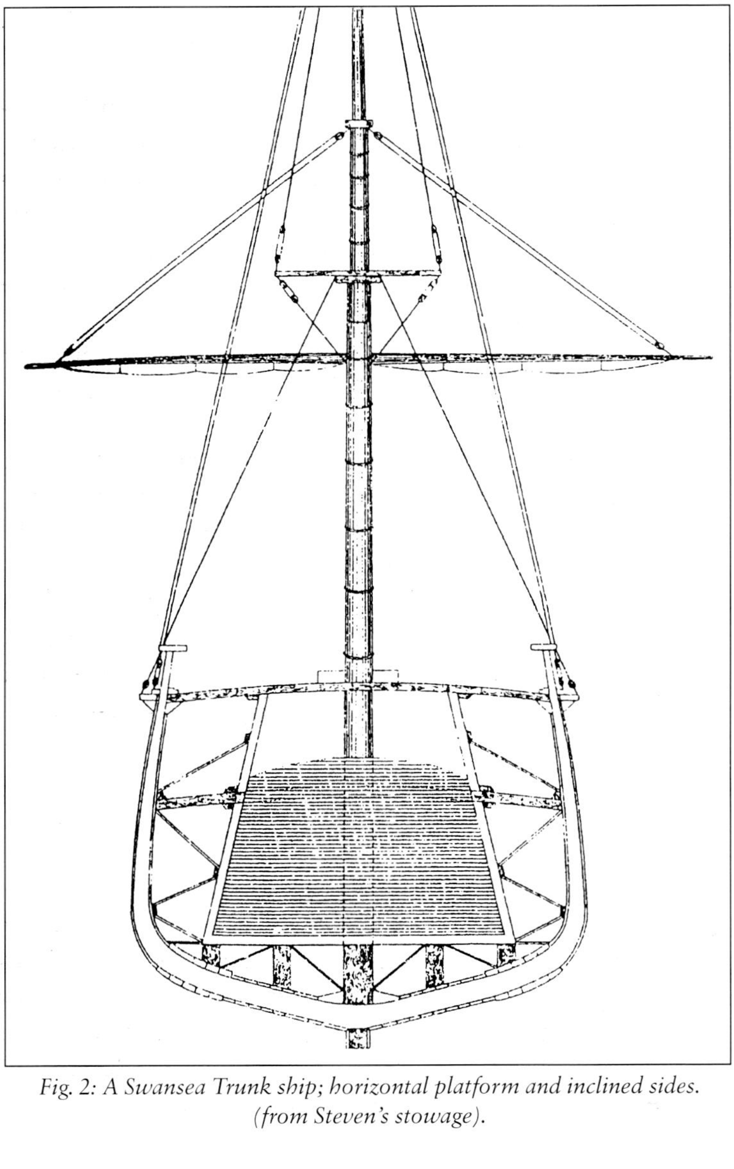

Fig. 2: A Swansea Trunk ship; horizontal platform and inclined sides. (from Steven's stowage).

The floor of the hold was known as the ceiling, and before the vessel was launched this would be fitted, and the sides of the hold lined with two and a half inch thick timber secured by six inch spikes.

Copper is a dense and consequently very heavy cargo, which meant that the hold was never physically full when the maximum amount of cargo was loaded. Stowed on the ceiling (the floor of the hold) it created a very low centre of gravity which gave rise to strains everywhere, including the masts and rigging.

To overcome this problem, a special trunking arrangement was devised, which came to be known as 'Swansea Fitted'. This consisted of a trunking assembly which kept the cargo away from the sides of the hold, and also raised the centre of gravity. It is best appreciated by referring to Fig. 2, page 96.

Making the sides of the trunking slope inward rendered the system even more effective. Such trunking was an expensive but necessary feature of the copper barques, and it was usually designed so that it could be dismantled if the vessel was required to carry more standard cargoes. Trunking also helped to prevent that ever present source of ship losses – cargo shifting. Such specialist techniques as this, combined with years of experience and understanding of the Cape Horn trade, resulted in many ships surviving a long time without accident.

An even more interesting example of the longevity of an iron sailing vessel was the *Langland*, built in Middlesborough by Raylton Dixon in 1875 for W. H. Tucker, of Swansea. For twenty years she was engaged in the Swansea/Chile copper trade before being sold to the Balestrino brothers of Genoa in 1897, and re-named *Maria Madre*.

In 1902 she arrived at Paysandu in Uruguay with a cargo of salt from Ibiza. An argument ensued between the master and a shipper over a cargo of hides destined for Europe, and the subsequent dispute resulted in the vessel spending the next 26 years laid up at Paysandu.

In 1928 she was sold to pay harbour dues and in 1943 engined and re-commissioned as a twin screw motor vessel. In 1948 she re-emerged as a three-masted schooner and in June 1957, while on passage from Santos to Montevideo ran ashore in fog and was pounded to pieces by heavy seas. She was a remarkable tribute to British shipbuilding.

IRON HULLS

A wooden vessel required constant pumping, and as it became older the difficulties in keeping it watertight steadily increased. Percy Lubbock, in his *Last of the Windjammers*, Vol. I, wrote: 'There used to be an old Swansea barque afloat a year or so before the war (World War I), the pumps on board of which bore silent testimony to their own importance, for their fly-wheel handles extended right across the main deck from pin rail to pin rail, and the bushes of the bearings were

The 'Maria Madre', ex 'Langland', of Swansea at Paysandu 1926, having lain there since 1902 (see page 97).
(National Maritime Museum, Greenwich)

so worn as to hint at long and continuous spells of pumping when every sea poured across the main deck, waist-high.'

An iron vessel, on the other hand, was a tight ship, and was almost indestructible, especially those built with the early thick iron plates. If not lost at sea they invariably ended as storage hulks or barges, and finally had to be broken up.

An early demonstration of the durability and strength of iron shipbuilding came from Brunel's steamer, the *Great Britain*, which went ashore on the coast of Ireland and remained there from September 1846 until August of the following year, continuously hammered by winter storms. John Grantham in his book *Iron Ship-Building* (London 1858), points out that: 'every part of the bottom had been battered and damaged by the rocks on which it was beaten, yet it was difficult to perceive the slightest strain on the upper sides or to detect any alteration in her form'. When eventually she was towed off, the hull remained watertight enough for her to float normally.

At first the kind of skills needed to build an iron ship were hard to find, because of the lack of precedent, and understandably not much help was forthcoming from the builders of wooden vessels, who naturally hated iron ones. The iron shipbuilders therefore had to develop from the race of iron-founders, boiler-makers, and blacksmiths.

Iron vessels had the following advantages over wooden construction:

1. Cheaper than a wooden vessel of the same capacity.
2. They were preferred by marine insurance companies in terms of risk.
3. They were infinitely more long-lasting.
4. They were cheaper to maintain.
5. They carried more cargo for the same tonnage.

One serious disadvantage of iron hulls was the difficulty of keeping them free of weeds and barnacles, which created such problems in speed and steering. Iron vessels trading in tropical waters had to be regularly dry-docked – at least once a year – in order to have tons of barnacles scraped off the bottom. Eventually, anti-fouling paints were developed which reduced the problem, but it was never totally overcome.

The main design changes resulting from the use of iron for sailing vessels was in the shape of the stem, which could be much more clean-cut, and the old transom counter was replaced by an elliptical one which possessed the virtue of considerable extra strength at a critical point.

Efforts to lay down rules governing the construction of iron vessels to ensure their safety met with little support from the builders in the early stages, for reasons of self-interest, and also because they themselves were still only finding their way. Lloyd's Register found it necessary to handle the registration of such vessels by tactful compromise, and the problem led to the Liverpool Underwriter's Association in 1862 issuing their own system for surveying and classifying iron ships.

Glasgow, Tyneside, and Liverpool were the first areas to commence building iron vessels in any numbers in the 1850s, but by the late 1860s iron shipbuilding had become general. In Swansea, shipowners tended to cling to wood for as long as they could, but even they began to buy in iron vessels.

Steel construction followed more slowly, because of its high cost, but by the 1880s, when improved methods of steel production made it cheaper, steel became more popular.

RIGGING AND DECK FITTINGS

The mass of cordage and chain which forms the rigging of a sailing vessel may appear overwhelmingly complicated to the lay person, but to a trained seaman every rope and belaying pin had its appointed place. It meant that an experienced seaman could go aboard any vessel, and even on the darkest night know exactly where to lay his hand on any particular rope.

The various types of sailing vessel are differentiated by their rigging, i.e. the number of masts, and the arrangement of the sails. A vessel with three or more masts and square rigged throughout is known as a ship.

A barque has three or more masts with fore and aft rig on the after mast and square rig on the remainder.

A barquentine has three or more masts, but only square rig on the foremast; the remainder are fore and aft rigged.

A brig is two-masted and fully square rigged.

A snow is similar to a brig, but has a triangular sail set between the masts with its lower fore-end running to a step on the deck.

A true brigantine has two masts, fully square rigged on the foremast. The after mast has a square rig sail at the top and fore and aft rig at the bottom. This rig fell out of use in the 1840s, being replaced by the hemaphrodite brig which has a square rigged fore mast and fore and aft rigged main mast, i.e. what is usually termed a brigantine.

A schooner is fore and aft rigged throughout. Where a square rig sail is carried on the foremast it is known as a top-sail schooner.

The brigantine, as the smallest type of square-rigged vessel, was popular because of course it was cheaper to build, and especially to rig. For example, the foremast of a four-masted barquentine would cost as much as the total of the other three masts.

Where a larger vessel was required the barque rig was more economical than that of a ship, and as the 19th century wore on it was used for larger and larger vessels, even up to 3,000 tons, especially in the nitrate and guano trades to the west coast of South America.

This may be an appropriate moment to clear two points of terminology which arise when writing about the sea and ships. The expression 'ship' as defined above is clearly inappropriate for barques, brigs, schooners, etc; the overall term 'vessel' covers all types of craft and therefore should be used where appropriate. However, 'ship' is so universally used by the general public I have given myself a dispensation in this respect. In any case Victorian captains of barques often, in fact usually described their craft as 'my ship', whatever its rig.

The same applies to the term 'captain' as the individual in command of a vessel. 'Captain' is strictly speaking a naval rank, next below that of rear-admiral, and its use in the Merchant Service is purely as a courtesy title; the correct term is 'Master', i.e. an individual qualified by examination to take command of a merchant vessel. Nevertheless, the term 'Captain' was used universally.

Though not part of the rigging, it may be convenient here to mention the painted ports, so popular on sailing vessels. The original idea was to give a vessel the appearance of being armed, and they became, so to speak, an established tradition which lasted as long as sailing vessels survived.

SAIL LOFTS

Sails were manufactured in special workshops known as sail lofts, and in the sailing ship era nearly every seaport town would have had at least one.

They normally had polished floors, and the sail-makers each sat on a low stool to ply his palm and needle. In later years heavy gauge sewing machines came into use, and sometimes women were employed to work on them.

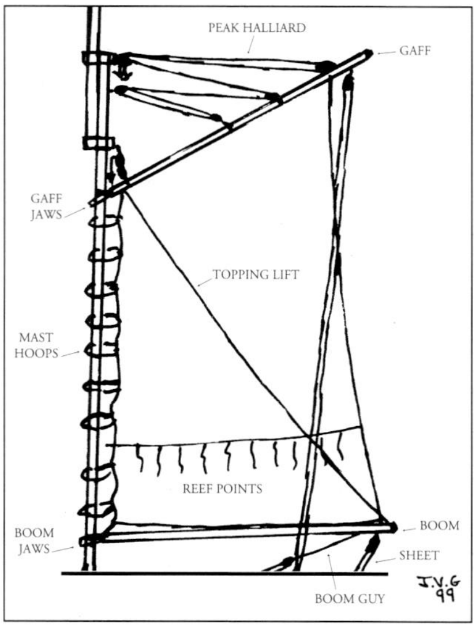

Fig. 3: Sketch plan of a fore and aft mainsail.

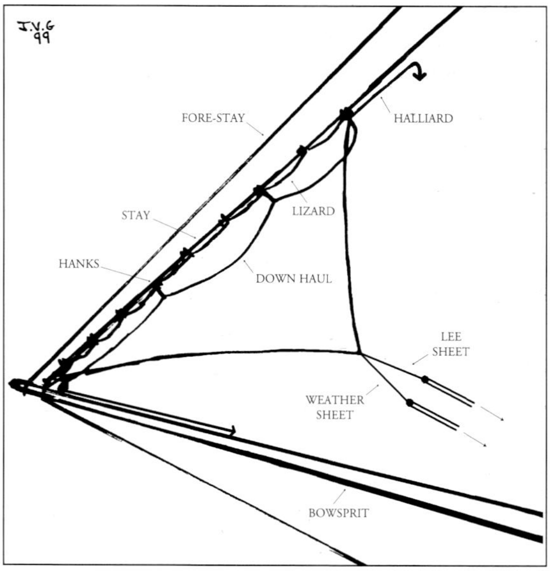

Fig. 4: Sketch plan of jib rigging and running gear.

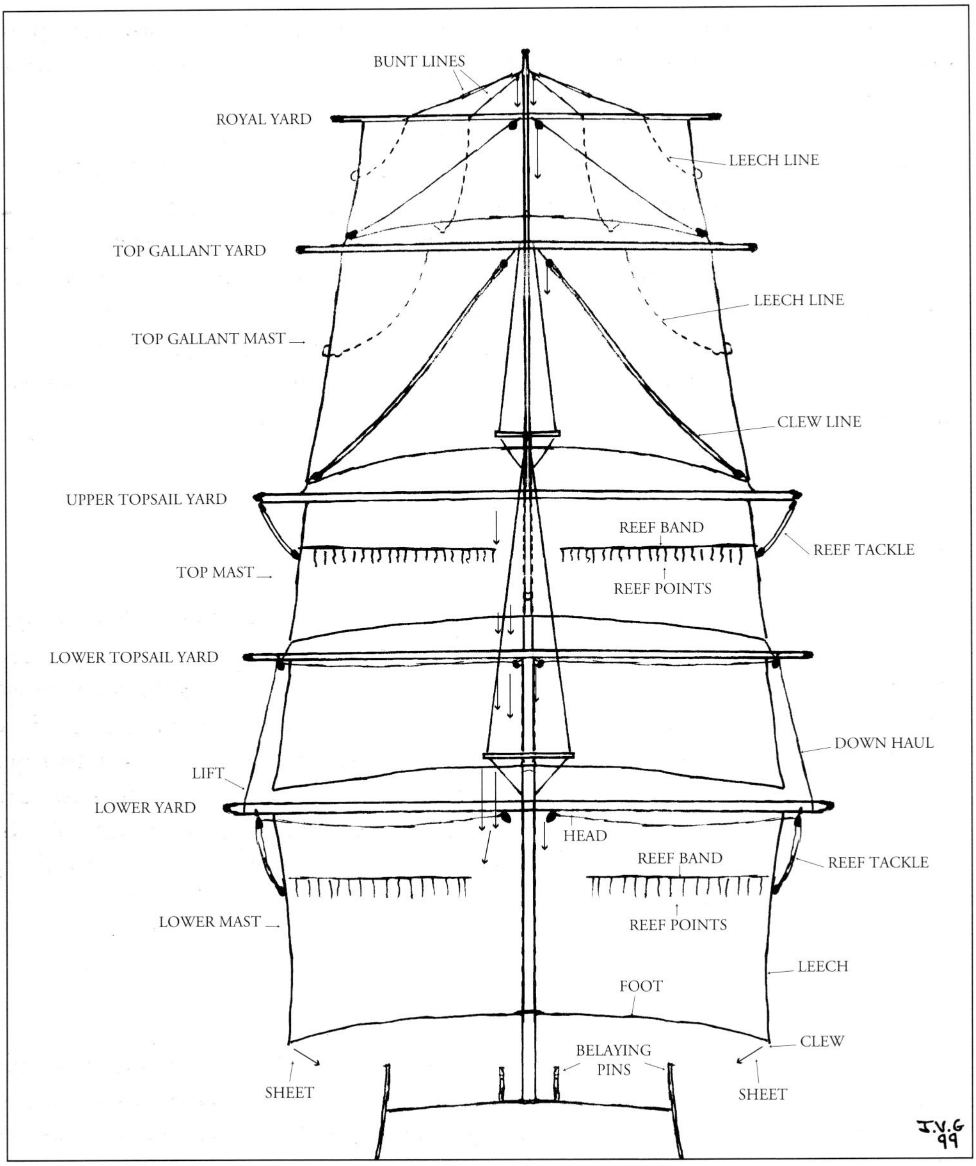

Fig. 5: Sketch plan showing square sails viewed from stern and main running gear. (Main sheets are only indicated).

The work consisted of making new sails, and of course repairing old ones. In charge would be a master sail-maker, and in the case of a new sail he would be provided with the mast and spar dimensions. He would then make a rough drawing of the sail outline on any piece of scrap cardboard that happened to be available. Some of these plans still exist, drawn in pencil on the backs of old calendars or posters. The card would show the cut of the sail and where all its constituent parts were to be sewn in; bolt ropes, reef-points, eyes, etc. (see Figs. 3, 4 and 5, pages 101, 102).

It would be the master sail-maker's responsibility to keep records of all the work done in the loft for charging purposes. He would also be expected to keep an account of the materials used, such as canvas, rope and tarred rope, oil, twine, thimbles, and so on. Sometimes, in repairing sails, second-hand canvas cut from old sails would be used to repair sails for customers who could not afford new material. The old sail lofts had their characteristic smells; that of new canvas, rope and tarred rope, and the distinctive odour of salt-soaked hemp rope, which no-one, including the present writer, can forget, once they have smelt it. (Though in her case it was mixed with steam.) Like the sound of caulking hammers in the shipyards, they were part of the life-blood of Victorian seaport towns.

ROPE WALKS

Ropes were hand-made, in a long building between 300 and 400 yards in length, where the yard was twisted into strands, and the strands twisted into rope. The best natural rope was made from hemp or manila fibre, which was pulled through boards studded with nails (called Hackleboards) to separate the fibre and get it to lie straight. The fibres were then spun into strands, which was laid up into rope. The strands were attached to the spinning machine and passed through a conical-shaped wooden former with grooves down it to keep the strands separate. As the strands were twisted into rope the rope maker had to walk down the length of the rope-walk moving the former, the final rope emerging from its thin end.

THE GALLEY

The galley, where food was cooked, was usually a shed-like structure normally situated behind the foremast, which provided it with some degree of shelter. It was sometimes a separate construction, secured to the deck by iron straps and bolts, so that it could be removed and secured above a deck cargo. Cooking was done on a coal-fired iron stove fitted with iron rails on the top to prevent the pots sliding off as the ship rolled.

DECK FITTINGS AND HOUSINGS

The standing rigging was secured at deck level by means of 'dead-eyes' (circular blocks of lignum vitae or elm), or rigging screws to the chain-plates, which were strips of iron or bronze whose lower ends were bolted to the ship's side. As well as the wash-ports and scuppers which have already been mentioned, the bulwarks had oval openings adjacent to the mooring bollards, which were fixed to the main deck for the purpose of securing the mooring ropes.

Sailing vessels always carried spare spars – usually two topmasts, and two more that could be made into yards. These were normally stowed on chocks on the main deck. Fife-rails were the wooden rails on the poop or main deck at the base of the masts which held the belaying pins used to secure the sail halliards.

The pumps, so vital in a wooden vessel, were situated on the main deck, with removable handles that could be stowed in special sockets nearby. Heavy flywheels helped to keep them working steadily as they were turned by hand.

Heavy lifting work was carried out by means of a capstan, which consisted of a cylindrical barrel vertically lined with ridges (called 'whelps') to give a grip for ropes or cables. Leverage to turn the capstan was provided by removable 'capstan bars' A windlass performed the same function, but using a horizontal barrel, again being turned by bars to provide leverage. In both case a pawl mechanism took the strain. A later variation allowed a 'pump handle' operation, based on the pawl system, but this was slower, and produced less torque than a capstan with bars, though that required the physical room for several men to operate it.

The steering gear consisted basically of a tiller fixed to the top of the rudder, which could be pulled in either direction by ropes or chains passing over a barrel arrangement connected to the steering wheel.

The steps or stairway leading down below from deck level were known as 'companions' and the aperture leading to them were covered by suitable waterproof covers, usually fitted with some kind of sliding hatch arrangement. Access to the holds was through hatches, which could be covered over with heavy planks, usually iron-bound at the edges, which were known as hatch-boards. Over these was stretched a heavy canvas cover, held round the hatch sides by iron bars secured by wooden wedges, driven in by a hammer.

This may be a suitable point at which to include some definitions of commonly used terms the general reader may encounter in this or other books on sailing ships:

Aback	When the sails of a square-rigged vessel are trimmed to catch the wind on the forward side. Sometimes this is done to slow the ship down, e.g. when about to anchor, or it can happen accidentally because of a sudden change in wind direction, or bad steering.
Alow	The opposite of aloft.
Braces	Wires or ropes attached to the ends of the yards in a square-rigged vessel, so that they can be trimmed.
Caboose	A wooden house generally on a wooden vessel, often the galley.

Clench	To make a permanent join.
Clew	An 'eye' (or 'thimble') stitched into a sail to take the ropes.
Clew-lines	The lines by which a sail is hauled up to the yard when it needs to be furled.
Dogs	Metal clips or screw bolts which hold doors or hatches in a closed position.
Drum head	The top of a capstan.
Fiddley	A grating in the deck to provide ventilation (sometimes access).
Gammon iron	An iron band which secures the bowsprit to the stem of a vessel.
Gammon lashing	The same as above, but done with rope.
Jack stay	A wire or rope between two points, to give support.
Jib	A triangular sail set between the bow and fore mast.
Nautical mile	Is different from a land mile. It is defined as the distance equivalent to one minute of latitude at the centre of the earth. Because the earth is not a perfect sphere, it varies with latitude. It is taken as approximately 6,080 feet.
Pampero	Is a violent squall (common off the coast of South America).
Rabbet	An incision in a piece of wood to receive the end or side of a plank.
Ratline	One of the 'steps' up the shrouds of a mast.
Sea anchor	Anything that will hold the bow of a vessel to the sea in heavy weather.
Sheet	A line used to trim a sail to the wind.
Shrouds	The standing rigging which provides a mast with its support.
Spar	A general term covering yards, booms, gaffs, etc.
Stay	A part of the standing rigging which supports a mast in the fore and aft line.
Waist	That part of the deck which lies between the forecastle and the quarter deck (i.e. between the fore and main masts).

8.

Life at Sea

Oft times to me come memories
Of fifty years ago,
Of Valparaiso – Carrizal
Coquimbo, Callao;
Of sweltering in the tropics,
Freezing off the Horn,
Pamperos off the River Plate,
Sails to ribbons torn.
Of running down the Easting,
Rolling in the Bay,
Soundings in the Channel,
(Counting up our pay)
Slush lights lit the forecastle –
A dark and dirty den;
Menu, salt horse, pantiles,
Wages two pounds ten.

Groping up the Channel
Light gleams on our lee;
Daybreak – Lundy Island,
First landfall to see.
Faces then were smiling,
Never a sneer or frown;
For we knew we'd shortly be
In 'Dear old Swansea Town'.
But age is ever creeping on,
My action's getting slow,
It wont be long ere eight bells strike
My final watch below!

A. J. Roman
Batum
Ex *Sea Breezes*. Quoted by
Capt. C. E. Mock, May 1939.

THE WORK – A TYPICAL VOYAGE

The crew would sign on ship's articles at the Board of Trade Shipping Office, where there would usually be a board outside, giving the names of the vessels signing off or on that day. The articles constituted a form of Agreement, which laid down the conditions of service, specifying wages, food allowance, the nature and duration of the voyage and so on (see Appendix IV).

When loaded, the vessel would be towed out to sea by a steam tug. If the anchors had been used in harbour these would have to be got on board, heaved up by a capstan which was turned manually by wooden bars.

The anchors would be stowed in cradles on the forecastle deck by small chains attached to ring-bolts. As soon as time permitted, they were usually coated with thick tar to prevent the salt spray from rusting them. The anchor chain was stowed

The crew of the 'Vanduara'.
(Swansea Maritime Museum)

in the chain locker, which lay below the crew's living quarters in the forecastle, and it would pass through their living area, coated with mud and silt from the harbour bed and radiating smells and damp.

On approaching port the process was reversed; the anchors would have to be swung outboard, and the chain got up from the anchor locker and shackled to the ring of each anchor by the carpenter.

On leaving Swansea the first job for the crew would be to clean off the coal dust and scrub down with a mixture of soda, ground pumice and saltwater known as 'sooji'. The next priority would be to check and if necessary overhaul the standing rigging, on which men's lives would depend. The shrouds (which support the mast on either side of the vessel), were fitted with ratlines which corresponded to the steps of a ladder for climbing up the mast. They were made of rope, or wood, or round iron bars; a faulty ratline could plunge a man to instant death. The other vital pieces of rigging were the foot ropes, which ran under each yard, on which a man stood to take in or set a sail. If one of these broke, a whole row of men working aloft might fall to their deaths.

The work of the crew was divided into periods of four hours, known as watches. This meant six four hour watches, beginning at midnight, but the watch between 1600 and 2000 was split into two, called the first and second dog watch. This enabled the crew to be divided into two watches with each watch alternating their periods of duty. The watches, commencing at midnight, were named middle, morning, fore-noon, afternoon, first dog, last dog, and first.

The passage of time during each watch was indicated by sounding a bell every half an hour, e.g. four bells struck at 1800 marked the end of the first dog watch; one bell at 1830 marked the end of the first half hour of the last dog watch. However, at 2000 eight bells were struck to mark the end of the last dog watch (1530 would be seven bells, 1300 two bells, etc.).

After crossing the Bay of Biscay and picking up the southerly trade winds there would be about two months of good weather, during which the rest of the rigging, and especially the running rigging would be inspected and overhauled where needed.

About 100 to 200 miles north and south of the equator were the Doldrums, where light winds and calms prevailed. It required a lot of hard work, shifting and trimming sails, to take advantage of every bit of breeze, and as the ropes were

A group of Swansea Cape Horners, circa 1920s.
(Captain Peter Rousey, 2nd from left in second row. See page 224).
(Swansea Maritime Museum)

usually sodden with rain and salt, men's hands became very sore with constant pulling.

South of the Doldrums conditions became more easy, but as the vessel approached Cape Horn, the heavy Cape Horn sails would have to be fitted, one at a time in turn, and this usually occupied at least a week.

The appalling conditions in the vicinity of the Cape Horn have been described in Chapter One. Once round the corner and into the Pacific the weather improved again until the vessel arrived at a discharging port in Chile, Peru, or San Francisco.

All the work of discharging cargo had to be carried out by the crew, some filling buckets, while others worked at the hand winches. Work was usually from 6 a.m. to 6 p.m. with threequarters of an hour for breakfast, and an hour for dinner. Saturday was a full day. But Sunday was always a day off.

The crew loaded the cargo in the same manner, either off lighters or from a wharf if the vessel was lying alongside. The work sequence on the voyage home was more or less the reverse of that on the passage out, and the next hazard, which all on board a Swansea-bound vessel viewed with apprehension, was the landfall in the Bristol Channel – for a sailing ship, one of the most dangerous stretches of water in the world.

A vessel may well have sailed all the way from Cape Horn without sighting land, and perhaps for four previous days before entering the Channel bad weather conditions could have made observations impossible. This would mean navigating by dead reckoning in an area notorious for contrary tides and treacherous currents, with rock studded coasts on either side. Sometimes a tug would come out from Milford Haven and offer a tow up to Swansea; then the Captain would have to decide whether to incur the high cost of towing to Swansea or manage without.

THE CREW HIERARCHY

The typical number of crew on a Swansea copper barque ranged from fifteen to twenty, though sometimes towards the end of the 19th century, when times became increasingly hard for sailing vessels, it could be as few as twelve. In command was the Master, with two officers, the Mate and Second Mate. Only large sailing ships carried a Third Mate.

Master

The Master did not keep a watch, so that the Mate or Second Mate worked a rota of four hours on, and four hours off. The Master of a sailing vessel was truly lord of all he surveyed, and whether it was a 'happy ship' or a miserable one, depended entirely on how he chose to run his vessel. His responsibilities covered a wide field, because as well as commanding the vessel, he had to safeguard the owner's commercial interests, which frequently included arranging for cargoes and bargaining

The barque 'John Williams', 1868.
(London Missionary Society)

over rates. In the days before radio and telegraphic communication, a Master would have to rely solely upon his judgement and experience. In the major ports there might be an agent employed by the shipowner, but in the end it was still the Master who had to take responsibility. Some shipowners allowed their Masters to carry cargo on their own account, provided there was space, and of course he would usually make sure there was some. Or he might be a part owner of the vessel he commanded, even if that only amounted to a few of the sixty-four shares into which the ownership of sailing vessels were usually divided. That gave him an extra interest in keeping the vessel in good order, and ensuring that running costs were kept low, sometimes to the crew's detriment.

The crew's wages were stated on the Ship's Agreement, which all signed at the beginning of a voyage, but the Master's salary was never shown. Fifteen, perhaps twenty, pounds a month was probably a general figure, but for the Master there were plenty of perks: when ordering stores aboard, fixing cargoes, perhaps selling old ropes and most of all, the standard of feeding aboard.

An unscrupulous Master in collusion with a cook/steward could make a good profit by buying the cheapest possible stores of meat and vegetables, and there was often a difference between the food eaten by the Master and officers, and the crew.

Fortunately, on Swansea vessels such Masters were the exception rather than the rule. However, they did exist, and some deliberately made their crews' lives miserable by calling them out in their watch below when it was not really necessary, providing uneatable food and generally bullying them with the deliberate intention of forcing them to desert when the vessel arrived in port. (In that case the wages due to them were forfeit and pocketed by the captain.) There is little doubt that some of the ships which were lost may have succumbed because the crews were exhausted, demoralised and underfed.

Another source of income was the 'Captain's Slop Chest'. This consisted of second-hand clothing which he would sell to members of the crew at a high profit, deducting the cost from their wages. Between slops, tobacco and disciplinary fines (if a sailor accidentally dropped a paintbrush overboard he would be charged for a new brush) it was not uncommon for a man to have very little pay due to him at the end of a voyage, especially if an allotment was being paid to his family at home.

Under the Mercantile Marine Act of 1850 the Master had absolute power to levy fines, dismiss offenders, or even put them in irons. The minimum fine was usually half a day's pay, and this could be given for some minor offence, such as swearing. All incidents on board were recorded in the ship's official log book, e.g. weather damage, accidents, injury, crew discipline and sometimes even attempted mutiny. The Master could shoot a member of the crew and get away with it, provided he had the support of his officers to prove an attempted mutiny. Even without it, a Court of Enquiry would probably give him the benefit of the doubt, such was the fear of mutiny in Victorian times.

Given this sort of background, it was hardly surprising that many of the more enterprising Masters managed to scrape together enough capital to become ship-owners themselves, and some like Walter Runciman went on to found great shipping companies.

Some Masters prided themselves on their ability to carry sail, an attribute not always appreciated by the crew, for occasionally it led to disaster. But most possessed superb seamanship capabilities and a weather lore that at times could seem almost uncanny.

One such character was Captain James Bevan, a Gower man who went to sea at the age of fifteen, retiring forty-three years later. During this period he went round Cape Horn fifty-six times (see page 126).

Mate

The mate was second in command; in the event of the Master not being able to continue in command, say because of accident or sickness, he automatically took charge by law. The minimum age of a mate was 19 with at least five years experience at sea, with one of those years as second mate. A typical wage would be £4.10.0 to £5.0.0 per month.

Second Mate

The second mate would take charge of a four hour watch in rotation with the mate. Minimum age was 17 with four years at sea. Typical wages £3.10.0 per month.

Boatswain (Bosun)

Supervised the day-to-day work of the crew, and, under the Mate was in charge generally of the maintenance of rigging, paintwork etc. Typical wages £2.10.0 to £3.00 per month.

Carpenter

The carpenter was responsible for the maintenance of all woodwork, boats, masts and spars, and the anchor and windlass. He was a key figure, and one of his most important tasks each day would be to check the level of bilge-water and the barrels containing food and fresh water. Typical wages £2.15.0 to £3.0.0 per month.

The carpenter was expected to provide his own box of tools, which were always his most prized possession. If through some mishap they were lost it would be a severe blow, for they were not covered by any kind of insurance.

Thomas Rodger, a ship's carpenter who was drowned at Guyayemil on 25th January 1861, was recorded in the vessel's official log as possessing the following list of tools, in addition to his personal effects:

14 caulking ions	2 axes
18 chisels	1 adze
8 gouges	1 draw knife
3 cold chisels	1 soldering iron
1 horsing iron	11 moulding and bead planes
1 stand saw	3 hand planes
1 circular saw	2 jack planes
2 drift bolts	1 pair pinchers
2 setting stones	2 pairs callipers
5 hammers	1 pair dividers
2 squares	4 small files
2 bevels	1 drill
1 screwdriver	3 bits
1 saw set	2 rules
a spoke-shave	

Able Seaman

Possessed the basic skills of steering, setting and reefing sails, and rope-work generally. Typical wages £2.10.0 per month.

Ordinary Seaman

Any man who served on deck as seaman. Typical wages £2.0.0. per month.

GRAVESTONES OF CAPE HORNERS

(Courtesy Mr Vivian Samuel, Swansea)

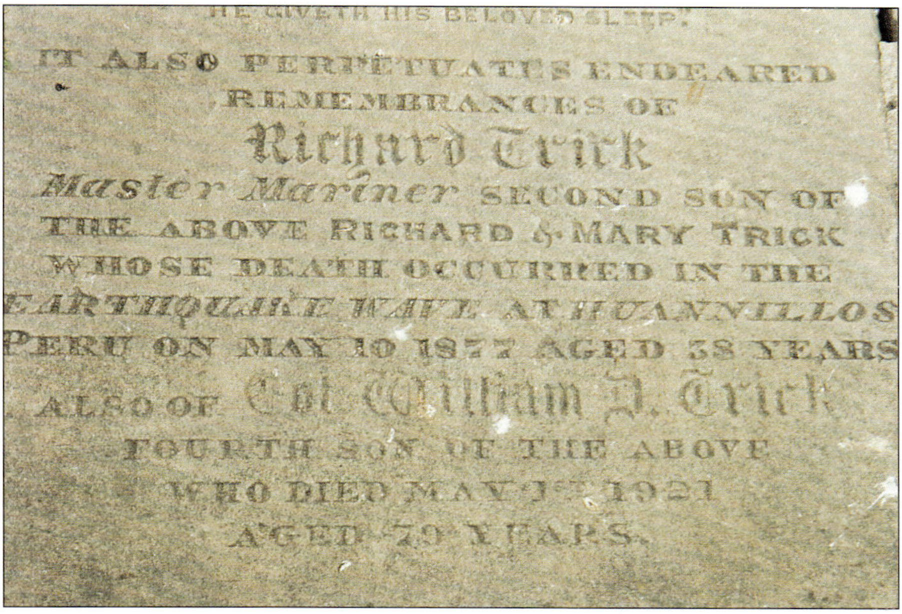

Richard Trick died in the earthquake wave at Huannillos, Peru, 10th May, 1877.

(St Mary's Parish Church, Swansea)

Llewelyn Jones, drowned while serving on the 'Kappa', 14th January, 1885.

(St Peter's Church, Cockett, Swansea)

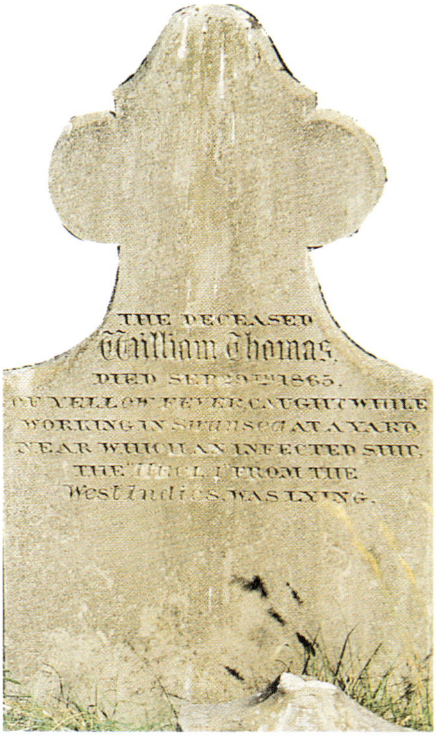

William Thomas, died of yellow fever brought from Cuba by the 'Hecla'.

(St Paul's Church, Sketty)

Mooring bollard.
Swansea Harbour Trust, 1842.
(Courtesy Mr Vivian Samuel, Swansea)

A ship's wheel (with author), 1999.
Maritime Museum, San Francisco.

The Cardiff Arms, Strand, Swansea.
(Courtesy Mr Vivian Samuel, Swansea)

Sail-maker
Attended to the sails and hatch and skylight covers and canvas work generally. He could mend sails, and if necessary, make them. If there was a death on board at sea, he would have the gruesome task of sewing up the body in canvas for burial at sea. Like the cook and the carpenter he did not stand regular watches. Typical wages were £2.15.0 per month. Sail-makers were generally only found on the larger vessels. On the barques such repairs would be carried out by an A.B., boatswain, or carpenter.

Stewards/Cooks
Cooks worked daylight hours and received no special training; they 'picked up' the simple cooking that prevailed on board. Quite often they were seamen who because of some injury were no longer able to go aloft.

Sometimes a steward was carried, who took charge of the food stores and waited at table on the officers, but this separate office was usually confined to larger vessels than the typical Swansea barques, where the two functions were often combined in one individual.

The cook would still be expected to help on deck when required, such as when wearing or tacking ship, or in emergencies. At such times his place would be on the foresheets, which were near the galley. Typical wage was £2.15.0.

Donkeyman
Some of the larger vessels in later years were fitted with small steam winches for discharging cargo, and the seaman selected to take charge of these was known as the donkeyman. (Steam engines were willing 'donkeys'.) Wages were from £2.10.0 to £2.15.0 per month.

Cabin Boys
Cabin boys went to sea at an early age, and were employed in assisting the cook and seamen. When they became big and strong enough they would be made ordinary seamen. Their pay would be little more than their board and keep and whatever pocket money the Master chose to give them.

Apprentices
There were usually at least two apprentices and incidentally provided the ship-owner with a cheap source of labour. They could begin as early as twelve years old, sometimes even earlier, to serve the four years required before they could sail as second mate. Their parents paid the shipping company a premium, usually of the order of £50–£75 per year. Their pay was purely nominal, perhaps 10/- per month, increasing to £1 in the last year of the apprenticeship, when they might act as 2nd Mate.

The old time seamen were masters of their craft, whether it was with a palm and

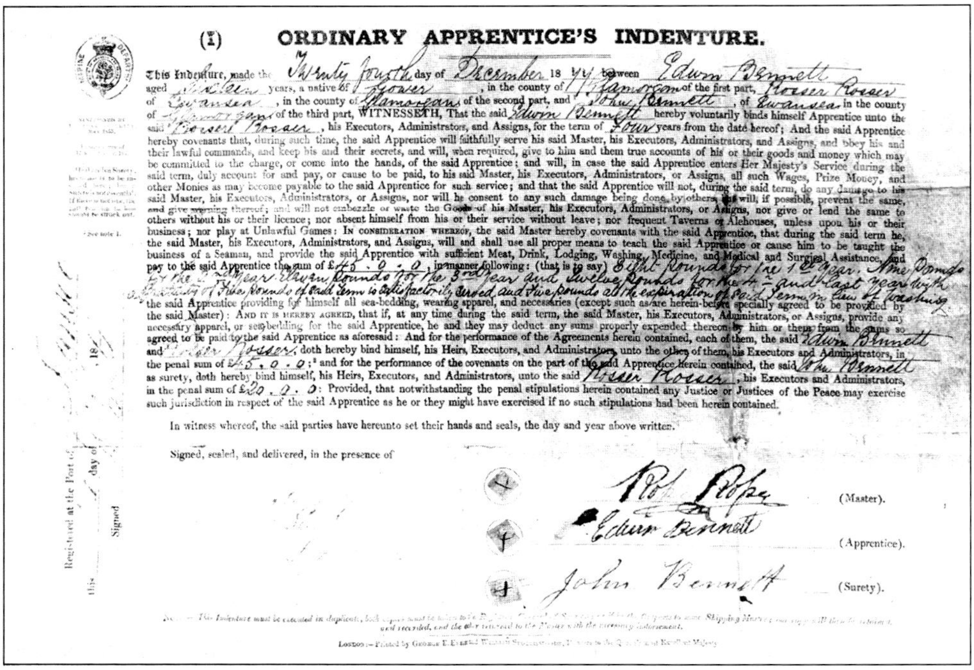

(I) **ORDINARY APPRENTICE'S INDENTURE.**

This Indenture, made the Twenty fourth day of December 1844 between Edwin Bennett aged sixteen years, a native of Gower, in the county of Glamorgan of the first part, Rosser Rosser of Swansea, in the county of Glamorgan of the second part, and John Bennett, of Swansea in the county of Glamorgan of the third part, WITNESSETH, That the said Edwin Bennett hereby voluntarily binds himself Apprentice unto the said Rosser Rosser, his Executors, Administrators, and Assigns, for the term of Four years from the date hereof; And the said Apprentice hereby covenants that, during such time, the said Apprentice will faithfully serve his said Master, his Executors, Administrators, and Assigns, and obey his and their lawful commands, and keep his and their secrets, and will, when required, give to him and them true accounts of his or their goods and money which may be committed to the charge, or come into the hands, of the said Apprentice; and will, in case the said Apprentice enters Her Majesty's Service during the said term, duly account for and pay, or cause to be paid, to his said Master, his Executors, Administrators, or Assigns, all such Wages, Prize Money, and other Monies as may become payable to the said Apprentice for such service; and that the said Apprentice will not, during the said term, do any damage to his said Master, his Executors, Administrators, or Assigns, nor will he consent to any such damage being done by others, but will, if possible, prevent the same, and give warning thereof; and will not embezzle or waste the Goods of his Master, his Executors, Administrators, or Assigns, nor give or lend the same to others without his or their licence; nor absent himself from his or their service without leave; nor frequent Taverns or Alehouses, unless upon his or their business; nor play at Unlawful Games: IN CONSIDERATION WHEREOF, the said Master hereby covenants with the said Apprentice, that during the said term he, the said Master, his Executors, Administrators, and Assigns, will and shall use all proper means to teach the said Apprentice or cause him to be taught the business of a Seaman, and provide the said Apprentice with sufficient Meat, Drink, Lodging, Washing, Medicine, and Medical and Surgical Assistance, and pay to the said Apprentice the sum of £45 . 0 . 0, in manner following: (that is to say) Eight Pounds for the 1st Year, Nine Pounds for the 2nd Year, Eleven Pounds for the 3rd Year and Twelve Pounds for the 4th and last year with Gratuity of Five Pounds if said Term is satisfactorily served and Five pounds at the expiration of said Term in lieu of washing the said Apprentice providing for himself all sea-bedding, wearing apparel, and necessaries (except such as are herein-before specially agreed to be provided by the said Master): AND IT IS HEREBY AGREED, that if, at any time during the said term, the said Master, his Executors, Administrators, or Assigns, provide any necessary apparel, or sea-bedding for the said Apprentice, he and they may deduct any sums properly expended thereon by him or them from the sums so agreed to be paid to the said Apprentice as aforesaid: And for the performance of the Agreements herein contained, each of them, the said Edwin Bennett and Rosser Rosser, doth hereby bind himself, his Heirs, Executors, and Administrators, unto the other of them, his Executors and Administrators, in the penal sum of £45 . 0 . 0; and for the performance of the covenants on the part of the said Apprentice herein contained, the said John Bennett as surety, doth hereby bind himself, his Heirs, Executors, and Administrators, unto the said Rosser Rosser, his Executors and Administrators, in the penal sum of £20 . 0 . 0: Provided, that notwithstanding the penal stipulations herein contained any Justice or Justices of the Peace may exercise such jurisdiction in respect of the said Apprentice as he or they might have exercised if no such stipulations had been herein contained.

In witness whereof, the said parties have hereunto set their hands and seals, the day and year above written.

Signed, sealed, and delivered, in the presence of

Robt. Rosser (Master).

Edwin Bennett (Apprentice).

John Bennett (Surety).

Registered at the Port of ... this ... day of ... 18.. Signed ...

Indenture of Edwin Bennett, Swansea.

(Courtesy Joe Johns, Swansea)

needle, marlin spike or adze. Usually they were very superstitious, and believed fervently in omens of good or bad luck, and all the antidotes to combat evil fortune.

A common belief was that, if they carried a caul round their neck in a small bag they would never drown, and this was a common practice. The story goes that one such bag was bought by a seaman from a Jewish junk dealer in London, but there was no caul actually in the bag. The man was later drowned at sea, and from then on if a seaman heard of some one drowning he would say 'he was wearing the Jew's caul'.

The sail-maker, boatswain, carpenter, cook and steward were frequently referred to as the 'idlers' because they were not on watches and consequently were able to get a full night's sleep. In reality, however, they worked just as hard as anyone else on board – probably harder for their daywork hours were long – at least from 6 a.m. to 6 p.m.

A young seaman was subjected to considerable peer pressure to learn his trade, and before an ordinary seaman was given a certificate of qualification as an Able Seaman he would have had to satisfy a shipboard tribunal consisting of the mate, boatswain, and perhaps an experienced A.B. Only then would the master award him the coveted title of A.B.

In his autobiography *Collier Brigs*, Sir Walter Runciman, who ran away to sea at the age of twelve, and commanded the barque *Althea* for four years before setting up as a shipowner, set down the tasks an A.B. was expected to perform:

1. Make short and long splices in small or thick ropes.
2. Put a clew and a cringle into either a fore and aft sail or a square sail.
3. Send any of the masts or yards up or down with ease and facility.
4. Graft with ease and facility.
5. Make bowline, Turks, whale and Rose knots.
6. Turn a deadeye in without a clumsy knuckle. Rattle down rigging neatly.
7. Make a carrick band with two ends of a hawser which the heaviest strain could never jam.
8. Rig a topmost stunsail boom.
9. Rig a top-gallant stunsail boom and set the sails.
10. Point rope ends to prevent them from tassling.
11. Stoop a double or single sheave-block.

A medical examination was not required for a seaman to join a vessel; to be able to walk on board, and produce satisfactory discharge papers from previous vessels was about all that was required. One could almost say that there was no intermediate stage between being fit for duty and dead.

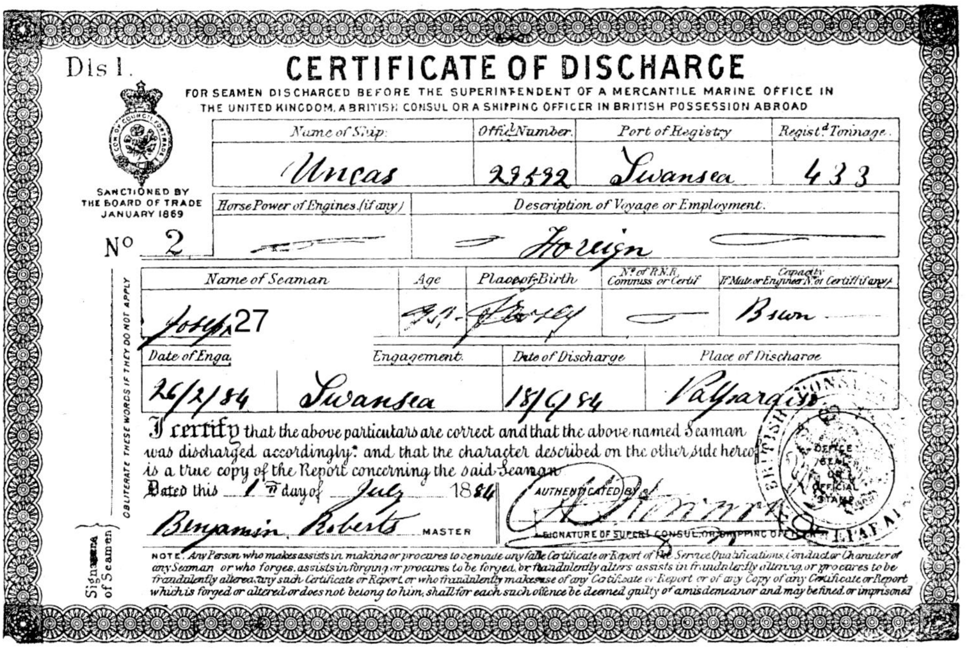
Dis 1.

CERTIFICATE OF DISCHARGE

FOR SEAMEN DISCHARGED BEFORE THE SUPERINTENDENT OF A MERCANTILE MARINE OFFICE IN THE UNITED KINGDOM, A BRITISH CONSUL OR A SHIPPING OFFICER IN BRITISH POSSESSION ABROAD

SANCTIONED BY THE BOARD OF TRADE JANUARY 1869

Nº 2

Name of Ship	Offl. Number	Port of Registry	Regist.d Tonnage
Uncas	29592	Swansea	433

Horse Power of Engines (if any)	Description of Voyage or Employment
—	Foreign

Name of Seaman	Age	Place of Birth	No. of R.N.R. Commiss. or Certif.	Capacity If Mate or Engineer No. of Certif. if any
Joseph 27			—	Bosun

Date of Engagement	Place of Engagement	Date of Discharge	Place of Discharge
26/2/84	Swansea	18/6/84	Valparaiso

I certify that the above particulars are correct and that the above named Seaman was discharged accordingly, and that the character described on the other side hereof is a true copy of the Report concerning the said Seaman.
Dated this 1st day of July 1884 AUTHENTICATED BY

Benjamin Roberts MASTER SIGNATURE OF SUPERT CONSUL OR SHIPPING OFFICER

OBLITERATE THESE WORDS IF THEY DO NOT APPLY

Signature of Seaman

NOTE. Any Person who makes assists in making or procures to be made any false Certificate or Report of the Service Qualifications, Conduct or Character of any Seaman or who forges, assists in forging or procures to be forged, or fraudulently alters, assists in fraudulently altering, or procures to be fraudulently altered any such Certificate or Report, or who fraudulently makes use of any Certificate or Report or of any Copy of any Certificate or Report which is forged or altered or does not belong to him, shall for each such offence be deemed guilty of a misdemeanor and may be fined or imprisoned.

Certificate of discharge of Joseph Nicholls, boatswain of the 'Uncas'.

(Courtesy Elvira Bevan, Swansea)

When a seaman died on board, his possessions might be pitifully small. Apart from the gear he, so to speak, stood up in, and a waterproof jacket of some kind, it might only amount to a couple of singlets, a few pairs of socks, a pair of blankets and often a book or a bible, or a pack of greasy playing cards.

Old seamen nearly always understated their age, sometimes by as much as twenty years, and many died on ships, of heart trouble, or simply being worn out, still claiming they were only fifty-five when probably they were nearer seventy. Sometimes a bottle of black hair dye would be discovered among an old A.B.'s kit after his death.

It is interesting to record that seamen generally took great pride in their appearance and Swansea men were particularly known for this. The standard rig consisted of blue dungarees, a clean jumper, white cap, often decorated by ornamental stitching, and small gold ear-rings, which all combined to create a distinct flair. Probably their romantic image and the famous vessels in which they sailed were a potent factor in attracting Swansea youths to follow the sea.

When a copper barque returned to Swansea after a voyage round Cape Horn it was an accepted practice for a relief crew to take over the vessel at the locks, so that the crew could go home immediately. The wages of the relief crew were paid by the crew; it was a purely unofficial arrangement condoned by the owners; they would sign off officially later. Another practice was for the crew to head to one of the public houses which lined the Strand and put a guinea on the bar counter – 'drinks for the house'.

It is pleasant at this point to quote an article from the *Cambrian* newspaper, 7th September 1860, entitled 'Laudable Mode of Rewarding a Ship's Crew':

> "On Wednesday evening, Capt. Paul Smith, of the fine ship *Marshal Pellissier* of this port recently arrived, belonging to Messrs. Richardson and Co., entertained the whole of his crew at the Duke Hotel where they were provided with an excellent and substantial dinner, served up in the host's best style. The chair was ably filled by Capt. Smith, the vice-chair by his chief officer Mr. Angrove, who made an excellent deputy. After ample justice had been done to the good things provided for the twenty noble young fellows, and the cloth removed, the Captain's health was proposed by Capt. Darling in eulogistic terms, and received with enthusiasm and drank with Swansea honours, as was also the health of the crew proposed by Capt. Smith, who explained to some of his friends who were invited at the festive board, that this dinner was given by him as a compliment to the men for their good conduct while laying in the port of Coquimbo, and during the voyage. When in that country, he promised that, on paying off in Swansea, he would not forget those who remained by the ship and did their work, instead of going ashore getting drunk and perhaps to

prison. On paying off the crew the other day, he offered them a pecuniary reward for good conduct; this the honest tars nobly declined, but would be glad to receive a mark of their captain's approval in some other form. It at once occurred to him to invite them to dinner, and was now glad to see so orderly and well-dressed a body of men respond to his invitation. The health of the officers of the ship, the excellent owners, and success to the *Marshal Pellissier* were equally well received, as was also the health of the host and hostess. After having spent a pleasant and agreeable evening, which was enlivened by song and sentiment, the party broke up highly delighted, and hoping to meet again on a similar occasion."

MASTER'S QUALIFICATIONS

The Merchant Shipping Acts of 1850 and 1854 introduced compulsory certification for Masters and the first certificates were valid for any kind of vessel; sail or steam, and every sort of rig. It was soon realised that the syllabus requirements excluded large numbers of experienced seamen who had no experience of square-rig; only of fore-and-aft, which was commonest. The Board of Trade consequently modified their regulations to issue these types of certificate:

Ordinary Master, who could command any vessel, sail or steam.
Master, Fore and Aft.
Master, Steam, who could not command a sailing vessel.

The new regulations resulted in a large number of vessels being rigged fore and aft, so that they could be commanded by masters with no experience of square-rig. The top sail schooner, which became so popular in the 19th century, was not classified as square-rigged as she only carried upper square-sails.

An interesting side-light on Masters' qualifications is given by the career of Sarah Jane Rees, of Llangrannog, Cardiganshire, in 1839. Her father commanded local vessels and she often went to sea with him on coastal voyages when he was short-handed. She qualified herself as a Navigator, and thereafter prepared local seamen for their Board of Trade examinations in navigation.

Sarah Jane Rees of Llangrannog, teacher of navigation.
(Courtesy Susan Campbell-Jones, Anglesey)

She won the crown for bardic poetry at the Eisteddford, and was also the founder of the Temperance Union of the Women of South Wales. She was a lay preacher, and founded and edited a women's journal in Welsh which ran for thirteen successful years. She died in 1916.[1]

CONDITIONS AT SEA

The captain and officers lived aft, and the crew in the forecastle, a dark and miserable place, accessible only through a hatchway on the deck down an eight or ten rung ladder. The hatchway, when the weather permitted it to be left open, provided the only light, except for an oil lamp, and often that was no more than old fat melted down over cotton wool in an empty corned-beef tin.

If there was heating, which was not always the case, it was provided by a small coal stove with a funnel fixed on top, going through the deck-head above. Food could not be cooked on it. The sides of the forecastle would be lined with bunks, and each man had to provide his own straw mattress or 'donkey's breakfast' as it was called. If they were lucky there might be a table. The only seating would be the men's own sea-chests. The apprentices, boatswain and carpenter would have their quarters in the half deck house.

In his reminiscences October 1954 (Swansea Maritime Museum M1.7197.2) Captain John E. Owen said, of life in the forecastle: '. . . there were old customs that had to be kept up. No sailor must lock his sea-chest. If he did so, it would be kicked open. The oldest seaman was the first to help himself to a meal, and it was an offence to eat with a cap on. Smoking on deck was forbidden abaft the foremast, and even an officer dare not smoke on duty. Cleaning up the living quarters was done by one man taking turns for a week at a time, except for once a week, when the watch below scrubbed out on a Saturday morning.'

The anchor cable came through the forecastle and as it was hauled in the links would be dripping with salt-water and rust scales and filled with mud; often stinking mud from the bottom of a harbour. In stormy weather as the vessel dipped her bows water would pour in through the chain holes, and flood the crew's quarters. Sometimes men slept in bunks with the water level just below them.

To quote from Captain Owen again: 'When lying at anchor, loading or discharging cargo on the west coast of South America, where there are no docks, the custom was that the crew could claim one day's liberty every six months. This was usually given when the homeward cargo was on board, and mostly on a Saturday night to Sunday night. They were advanced some pay and taken ashore in the ship's boat, one watch at a time. It wasn't often that they got far from the first dance hall, where they were welcomed by the ladies and gentlemen of the crimping profession, who soon got their money after making them drunk, and often enough beat them up as well.'

1. Susan Campbell-Jones, *Welsh Sail.*

Aboard ship. Drawing by James Harris Jnr.
(Private Collection)

THE PLIMSOLL LINE

The Swansea barques were renowned for their seaworthiness and the seamanship of their crews. Other ports were not so fortunate, for some unscrupulous shipowners deliberately sent ill-found ships to sea, overloaded and over-insured, literally hoping that the vessels would be lost. And occasionally there might even be collusion between the master and owner to fake a loss in some area where the crew could safely take to the boats and expect to be picked up.

The man who changed all this was Samuel Plimsoll, born in 1824. He began life as a clerk in a brewery and ended up as a successful coal merchant and a Member of Parliament. He dedicated his life and his health to his 'Plimsoll Bill' of 1871 calling for all vessels unclassed at Lloyds or by the Liverpool underwriters to be compulsorily surveyed and marked with a maximum local line.

The bill was largely ignored by Gladstone's government, and Plimsoll himself was savagely attacked by the shipowning lobby, who accused him of rendering the British Merchant Service non-profitable. There followed a long saga, and it was not until 1876, when Plimsoll had ceased to be a Member of Parliament, that an Act was passed giving the Board of Trade powers to require and police 'the Plimsoll Load Line on all except small local craft.'

When Plimsoll died in 1898 a group of seamen from the Chatham area interrupted the funeral procession, disconnected the horses from the shafts, and dragged the hearse to the cemetery, as a mark of the universal respect felt by all seamen to his name and memory.

There were other improvements; the Merchant Shipping Act of 1867 required lime or lemon juice to be carried on all foreign-going ships except those bound for European ports and the east coast of North America. The juice had to be of required strength, and come from bonded warehouses. Another Act in 1871 allowed men charged with desertion or refusing to sail because the vessel concerned was unseaworthy, to demand a survey, provided that at least a quarter of the crew supported the claim.

A major advance against 'crimping' came in 1880 with the Merchant Seamen (Payment of Wages and Rating) Act. This made 'Advance Notes' illegal. Penalties were imposed on persons being on board without permission – designed to prevent crimps having access to easily persuaded seamen. Finally, imprisonment for desertion was abolished.

MISSIONS TO SEAMEN

The need to provide some way for seamen ashore, either out of work or off duty, was filled by the many missions to seamen that came to be established in the major ports of the world. Largely due to William Henry Giles Kingston, a Victorian businessman with a religious bent, the 'Society for Promoting Missions to Seamen Afloat at Home and Abroad' was established in April 1856 with him as honorary

The Admiralty frigate 'Thisbe', which served as Missions to Seamen Church and Institute in Cardiff Docks from 1863 to 1891.
(Mary L. Walrond: *Launching Out Into the Deep*)

A captain's wife feeds her chickens on board ship.
(Courtesy San Francisco Maritime Museum)

secretary, and Lord Shaftesbury, the leading Christian Layman of his day, as President. At the first public meeting in March 1857 Kingston could report that five chaplains were already at work in 'the Mersey, the Bristol Channel, the Isle of Wight, Milford and Swansea'.[2]

FOOD

The galley was usually a shed-type structure with sliding doors on either side. Salt water was invariably used for washing dishes, the fresh water being carried either in galvanised tanks on deck or stored in water-barrels.

The food stores were kept aft in wooden zinc lined cupboards, where they were safely under the eye of the Master. The cook often played a dual role as steward, and the degree of latitude and trust extended towards him by the Master would

2. Roald Kverndal, *Seamen's Missions; Their Origin and Early Growth*. Published by William Carey Library, Pasadena, USA.

A group of Llanelly Cape Horners.
(Private Collection)

depend on the relationship between them. Some Masters carried the same cook/steward for years: a combination which for the crew could have either have good or bad results. The amount of stores carried would be calculated for an average voyage, and it was an important part of the cook/steward's duties to spread the consumption of stores accordingly.

Rounding Cape Horn could sometimes add several weeks to the voyage (it was not unknown for the barques to sail right round the world in the opposite direction to get into the Pacific). So supplies of food and water would then have to be severely reduced, from what was already a bare minimum.

Food consisted basically of salted meat and ship's biscuit supplemented by oatmeal bread, potatoes, peas and rice, until supplies of these ran out. Salted meat or pork was packed in wooden barrels and was only eaten because the alternative was starvation. It smelt and tasted terrible. Ship's biscuit was a thick brick-like solid, made of flour and water, and was usually filled with weevils, which had to be tapped out. Bread was a special luxury and only provided when the weather was good enough for the cook to bake it, and for the limited period he had flour available.

Cooks usually made very good soups and dumplings. Sea pie, another favourite, consisted of layers of beef or pork, potatoes and onions separated by a pastry made up of pounded ship's biscuit. 'Burgoo' was the name given to a 'gruel-like' concoction made of boiled oatmeal, seasoned with butter, salt and sugar. Another staple dish was 'lobscouse', made from chopped up salt-meat, broken biscuit, potatoes, onion and salt boiled up into a stew.

The *Llanelly and County Guardian* of 15th September 1932 published an account of a gathering of forty-two retired Cape Horners at the Thomas Arms Hotel. The dinner menu is eloquently redolent of their memories of sea-going fare, for it included such names as Salt Horse, Harriet Lane (the sailor's term for meat, after

Cape Horners of Carmarthenshire

"Red Jacket" amongst the ice off Cape Horn August, 1854.

THE FIRST

Annual Re-union Dinner

AT THE

Thomas Arms Hotel, Llanelly,

On SATURDAY, SEPTEMBER 10th, 1932,

at 6-30 p m.

Chairman - - - - Capt. JOHN HOWELL.

Vice-Chairman - - Capt. EDWARD MORGAN.

Tea Clipper, "Cutty Sark."

The Sailing Cape Horners.

Capt. JOHN HOWELL, Llanelly.
" GEO. RICHARDS, "
" JOHN WILLIAMS, "
" ELIAS JONES, "
" EDWARD MORGAN, "
" HARRY EDWARDS, "
" T. REES, "
" R. MARKER, "
" R. MERCER, "
" E. HUMPHREYS, "
" GEO. T. CAINAN, "
Pilot W. HUGHES, "
" W. J. PHILLIPS, "
" W. J. CAINAN, "
" C. W. DARLING, "
H. ELLIS POOLE, "
TOM GRAVELLE, "
T. H. EVANS, "
BEN PERROTT "
CHR. WILLIAMS, "
G. J. HOPKINS, "
E. GAMMAGE, Llanelly.
D. HOPKINS, "
J. REES, "
T. HOPKINS, "
J. SAUNDERS, "
A. ROWLANDS, "
W. D. VIVIAN, "
D. BOWEN, "
T. H. EVANS, Burry Port.
W. H. MEXSOM, "
JOHN JONES, "
T. WHEATLAND, Pwll.
J. DAVIES, "
T. PHILLIPS, Ferryside.
E. JONES, Pontyeates.
H. JONES, Croesyceiliog.
JAMES JONES, Garnant.
F. GRIFFITHS, Llanstephan.
W. A. THOMAS, Aberavon.
MARK MORRIS, Llangennech.

MENU.

Soup and Bouilli.

Dolphin, Flying Fish, Benita's.

Cape Pigeons, Albatross.

Salt Horse, Harriet Lane,
Salt Pork and Molasses.

Seaweed, Yams, Boston Beans.

Dandy Funk, Dog's Body,
Rice and Molasses.

Cracker Hash.

Swankey and Lime Juice.

Toast List and Programme.

"The King."

Roll of Honour.

Sea Shanty.......... OLD CAPE HORNERS

"Old Cape Horners."

Proposed by H. W. SPOWART, Esq.

Recitation.......... Mr. TOM GRIFFITHS

Responded to by H. ELLIS POOLE, Esq.

Song.......... Mr. D. BRAZELL, A.R.A.M.

"The Trade of our Town and County."

Proposed by HIS WORSHIP THE MAYOR.

Song.......... Mr. J. AUCKLAND

Responded to by J. H. MONTGOMERY, Esq.

"Our Visitors."

Proposed by J. BOWEN, Esq.

Song.......... Mr. D. BRAZELL, A.R.A.M.

Responded to by Ald. MORGAN MORGAN, J.P.

Selected item.......... Mr. T. GRIFFITHS

PERSONAL REMINISCENCES.

Selected item.......... Mr. J. AUCKLAND

"Hên Wlad fy Nhadau."

"God Save the King."

Programme of Carmarthenshire Cape Horners Re-union Dinner at the Thomas Arms Hotel, Llanelly.
(Private Collection)

Jack the Ripper's exploits in London), Sea Weed Yams, Dog's Body, Cracker Hash (pounded biscuit and molasses), Boston Beans, Salt Pork and Molasses, and Dandy Funk.

Who better could there be than Captain John E. Owen (see page 146) to relate the scale of provisions as laid down by the ship's Articles of Agreement:

> Per week:
> 21 quarts of water for all purposes, (and the cook claimed most of it for soup), tea and coffee. Salt meat, fish and bread were cooked in salt water. Soft bread 3 lbs, biscuits 4 lbs, salt pork 3 lbs, tinned meat 2 lbs. Salt fish $^{3}/_{4}$ lb, potatoes 6 lbs as long as they lasted, dried vegetables $^{1}/_{2}$ lb, split peas $^{2}/_{3}$ pint, green peas $^{1}/_{3}$ pint, flour 1 lb, rice $^{1}/_{2}$ lb, oatmeal $^{1}/_{2}$ lb, tea $1^{3}/_{4}$ oz, coffee 4 ozs, sugar 14 ozs, butter $^{1}/_{2}$ lb, marmalade 1 lb, dried fruit 5 ozs, and, curry powder $^{1}/_{4}$ oz.

There is a story that on one vessel, the newly joined 2nd mate was about to spread butter on a ship's biscuit when the captain shouted at him to turn it over. It seems that he was spreading the butter on the side with weevil holes in it, which would soak up more than the prescribed quantity of butter. But happily on Swansea vessels such tales were few and far between.

SICKNESS OR ACCIDENTS

Lime juice was always carried and issued on a regular basis to prevent scurvy. It was very strong and not at all a pleasant drink taken on its own. It was first made mandatory on British vessels (hence the American term 'limeys') in the light of a report by Admiral Ryder submitted to the Society for Improving the Conditions of Merchant Seamen.

Sickness or accidents on board were often terrible events. A man's recovery from anything serious would be entirely dependent of the Captain or Mate's knowledge of rough medicine, derived from *The Ship Captain's Medical Guide*, a 1/6d manual carried on every vessel, and any clinical experience picked up over the years. First aid supplies were of the most basic kind; splints, bandages, and simple remedies like iodine, ipecacuanha, quinine, and Friar's Balsam.

The following list of mixtures extracted from *The Ship Captain's Medical Guide*, 14th Edition, published 1906 by Simpkin Marshall, Hamilton Kent & Co., London, gives a good idea of what was available. Some of the remedies would no doubt have been very effective on malingerers!

Fever Mixture

Nitrate of potash	1 drachm
Sweet spirits of nitre	3 drachms

Spirits of chloroform	1½ drachm
Add water to	6 oz

2 tablespoons for a dose.

Diarrhoea Mixture

Elixir of vitriol	2 drachms
Laudanum	½ drachm
Spirits of chloroform	1½ drachm
Add water to	6 oz

2 tablespoons for a dose.

Cough Mixture

(To be well shaken)

Powdered ipecacuanha	12 grains
Paregoric	2 drachms
Sweet spirits of nitre	3 drachms
Spirits of chloroform	2 drachms
Add water to	6 oz

2 tablespoonfuls for a dose.

Clap Mixture

(This must be well shaken).

Balsam of copaiba	3 drachms
Sweet spirits of nitre	2 drachms
Friar's balsam	1 drachm
Add water to	6 oz

2 tablespoons for a dose.

Stomachic Mixture

Spirits of chloroform	2 drachms
Bicarbonate of soda	2 drachms
Essence of ginger	1 drachm
Add water to	6 oz

2 tablespoonfuls for a dose.

9.

Local Cape Horners

THERE IS LITTLE DOUBT THAT Swansea was the home of more Cape Horners than any other port in the world. No doubt the ports on the north-eastern seaboard of the United States, homes of the 'Down Easters', would collectively rival it, but they are several while Swansea is one.

In researching this book I have been overwhelmed with the names of local Cape Horners, nearly all the great-grandfathers of people still living. Sadly, in only a few cases do memorabilia and particularly photographs survive. Even so, it has been difficult to choose, and those which follow are only a selection.

EDWIN BENNETT

Edwin Bennett was born at the village of Ilston in Gower, and at the age of 14 was appointed as a joiner to a firm in Swansea. He walked there on Monday and back on a Friday, lodging in Swansea during the week. At the age of 17 he literally downed tools and joined the *Erycina*, a 544-ton wooden barque owned by Rosser and Rosser of Swansea for a voyage to Chile. On his return he was apprenticed, and I am indebted to his grandson, Mr. John, of Swansea, for the reproduction of his indenture which appears on page 114, and his accompanying photograph.

After completing his apprenticeship he served as Second Mate, doubling Cape Horn at least four times before leaving the sea at the age of 22 to join the Customs Service, from which he retired after 30 years service.

The *Erycina* was lost en route from South Carolina to Birkenhead in December 1878. Edwin Bennett died in August 1953 at the age of 91; another example of the longevity of those Cape Horners who survived the Horn.

CAPTAIN JAMES BEVAN

Captain James Bevan was, with Captain David Morgan of the *Serena*, one of the most widely known and respected of Swansea sailing shipmasters. He was born in 1851, and went to sea at the age of fifteen, becoming master in his early twenties. His commands form a list of some of Swansea's most famous vessels, including the *Agnes Lillian*, *Brynymor*, *Pembroke Castle*, *Foxhound*, *Kandahar* and *Vanduara*, and he rounded Cape Horn fifty-six times without losing a ship. He was made Commodore of the Goldberg fleet, but steadfastly refused to go into steamships, preferring instead to retire at the age of fifty-eight to his beloved Gower.

Captain James, as he was universally known, was of medium build but very stocky, and his descendants, as I write in 1999, still have vivid memories of him. It was said that 'he had hands like shovels, and fingers like sausages'.

His grand-daughter, Phyllis Jones, now of Anglesey, has many anecdotes about him; he was a great walker and if the weather prevented him from going out (and it must have been bad) he would walk between his sitting-room and parlour, an activity known to the family as 'Grandfather walking the poop'. Sometimes he would give Phyllis a silver 3d piece 'to spend at Miss Taylor's shop'. In 1875 he married Margaret Bevan of Paviland; they had four sons and lived at Pembroke Cottage, Horton.

Captain and Mrs Bevan had their own pew in the Chapel at Horton – third on the right, while behind them always sat his brother, John Bevan, of Bank Farm, Gower. During the long years of his retirement he invested in property, and helped his sons set up in business, some of which still carry on today. One of these was the Pioneer Bus Company in 1911, which provided the first bus service in Gower.

A full history of his career is worthy of a book on its own. On one grim voyage to South America he lost all his crew to Yellow Fever, and the fever even claimed half the crew sent out to Santos in Brazil to replace them. On another occasion the *Brynymor*, owned by Hyam Goldberg of Swansea, sailed on her maiden voyage in November 1902 with coal for Port Elizabeth under the command of Captain James. It is an interesting commentary on the decline of the sailing ship era that his first and second mates were Captains Gregory and Potts. The vessel was caught in a storm in Bay of Biscay, which damaged the masts and sails and made it necessary for Captain James to return to Swansea. During the storm, Captain Potts was washed overboard by one sea and back by another, surviving with a broken arm.

When Margaret Bevan died on 1932, Captain James went to live with his sons in Swansea. He died in August 1938, and lies buried with his wife in Port Eynon churchyard. His picture appears on page 128.

I cannot end this account without expressing my thanks to his descendants, Phyllis Kelson Jones of Anglesey, Raymond Bevan of Worcester, and Michael Bevan of Swansea.

GEORGE DAVIES OF HAFOD, SWANSEA

Mrs. Mary Pope of Penclawdd has told me of her maternal grandfather, George Davies, who was born in the Hafod in January 1870 and went to sea at the age of thirteen as a ship's apprentice in the copper ore trade.

The vessel's cargo of coal caught fire somewhere in the vicinity of Tierra del

Edwin Bennett.
See indentures, page 114.
(Courtesy Joe Johns, Swansea)

Captain James Bevan outside his home, Pembroke Cottage, Horton, circa 1930.
(Courtesy Michael Bevan, Swansea)

George Davies.
(Courtesy Mary Pope, Penclawdd)

Captain Richard Evans.
(Courtesy Capt. Colin Evans, Swansea)

Captain John Jones.
Circa 1910.
(Courtesy Gwladys Jones, Swansea)

Captain Thomas Gwynne Evans.
(Courtesy Miss J. M. Evans, Burry Port)

Andrew Nelson.
(Courtesy Cyril Nelson, Swansea)

Joseph Nicholls.
Sailmaker and boatswain.
(Courtesy Elvira Bevan, Swansea)

Captain J. E. Owen.
(Swansea Maritime Museum)

Three masted barque off the Gower coast, by James Harris.
(Private Collection)

Shipping off the Mumbles, by James Harris.
(Private Collection)

Ships off Mumbles, circa 1870, by James Harris.
(Swansea Museum Collection)

'Illimani' off Mumbles, by Alexander K. Brander.
(Private Collection)

Fuego and to save his ship the captain beached her, so that the fire could be put out, the vessel refloated, and the cargo re-loaded. It seems that the crew became friendly with the local Indians, one of whom returned to Swansea on the ship, and later went back to his own country as a missionary.

By then the copper ore trade was diminishing, so George Davies, who by then had risen to second mate sailed on other ships to many parts of the world, including the famous clipper *Themopylae*.

Later in his career, while on a voyage from India to South Africa with immigrants, George felt that the Indian passengers were being badly fed and treated, and confronted the captain about it. He was accused of mutiny, and placed in irons until the vessel reached Liverpool, where he was given a bad discharge without pay. This effectively ruined his career at sea, and after a few years on coasters he left the sea to work on the Swansea tramways, and later as a docker. He died in May 1941.

CAPTAIN RICHARD EVANS

I am indebted to Captain Colin Evans of Swansea for the following extracts from an account written by his great uncle, Captain Richard Evans, of a voyage to Chile in the barque *Sinbad*.

A Midsummers Nights Dream
by Captain Richard Evans (1871-1948)

> The Barque *Sinbad* of Swansea, 615 tons N.R., owned by the Swansea Merchant Shipping Company, Gloucester Place, was I believe the last sailing ship to load in the North Dock for Chile.
>
> She was under the command of Capt R. W. Brisco and carried a crew of 16 hands all told. I, then nearing 24 years of age, was making my second voyage as First Mate in her and the men were a good type of the regular copper ore men, all Swansea men except one, the cook, who I believe hailed from Middlesborough . . .
>
> We loaded at Bath's yard with supplies for the mines at Copiapo, said cargo consisting of fire bricks, slabs about 35-40 sorts, shapes and sizes, cases of machinery, bolts, nuts etc, some pig iron, 300 tons of iron rails and finished off with coke and some smithy coal . . .
>
> The crew were signed on March 16th 1895 and owing to a long spell of westerly winds we did not sail until April 1st for Caldera, the port for Copiapo. I think all the old shellbacks were down to see us sail on that Monday morning and were full of talk that the palmy days of the Swansea copper ore trade were to come again.
>
> I think it interesting to comment on the mentality of the seamen of those days, when on the voyage outward the language used to express

their opinions of what should happen to them should they be found foolish enough to ship in a copper oreman again, was picturesque and forcibly expressive, but generally, when paid off, unless they were married, very few days would see them on their beam ends and away they'd go for another voyage in the same direction and in dirty weather while 'standing by' (for sailing) the refrain of their song would be 'and when my money is all spent, here's around Cape Horn for more' . . .

The *Sinbad* had a very fine figure head. It was built of oak and represented a burly figure of about 6 feet in height and was wonderfully carved; it was 'some job' painting it to our satisfaction but we used to manage a fairly gorgeous reproduction of the original in his robes. The diamonds in his turban used to beat us so sometimes *Sinbad* wore rubies and at others sapphires in his turban. Anyhow as he had been surf bashing around the Horn for 20 years he was pretty well seasoned and the bolts holding him in position had rusted and the surrounding timber had rotted so he was not holding a very secure position. One bright, and for a wonder, sunny morning he took his departure when the ship put her bows under and it was more in the nature of the sergeant-majors farewell, because in leaving us he carried away the lee back-ropes and of course all the head gear was adrift. Another Sunday job fixing the damage, taking it in turns to work one or two men holding on to the worker to prevent him from being washed overboard and another to hold the marline spike between times, as, if it was placed on the deck, it promptly got frozen fast in spite of the bright sunshine, and altogether we did not think that *Sinbad* would have become our old man of the sea as he did . . .

The German and French ships were always cracked up as being the ships to make the best passage around the Horn but there can be no comparison between the ships. Just fancy any English, I mean (Great) British owner providing bunker coal to make the sailors task lighter. I know for certain that Messrs A. D. Bordes' vessels used to have steam on the donkey (engines) as soon as they got to 40°S. Latitude and if a lull (in the wind) came, up went the topsails, by steam of course and the captain found himself on the carpet if his log showed that the ship had stood on too far south. This of course was the common sense way but our owners thought more of the sixpence they had to spend instead of the shilling they could have saved and further, in our excuse (of) the 'English' mostly, the mate was expected to 'pinch' enough coal to run his donkey (engine). I know of at least one acquired Swansea ship, a big one, fitted with steam, but all her cargo of saltpetre was loaded by hand winch because there was no fuel for the donkey.

We duly arrived at Caldera on a Friday afternoon and berthed at the wharf on Saturday afternoon and our band went down to the end of the wharf to play out the Pacific Steam Navigation Company's steamer *Inca*, bound for Swansea. The steamer, even then, was cutting out the small copper ore ships and as a decent sized steamer, then about 3,000 tons D.W. (Dead-weight), could take an ordinary copper oreman's cargo as part cargo, without the expense of building the elaborate trunk stringered and shored from the ship's side with three rows of shores or 'Toms' (as) we called them. The apprentices were the trunk building squad as they made good (i.e. long) stays (i.e. periods) in each ship (and) sometimes served the whole period of their indentures in the same ship and knew each trunk plank by sight. I know I did. And many's the jammed finger I got when my pal on the 'tween deck beams or vice-versa, did not close the planks in unison. 3 inch x 8 inch deals were used.

In Caldera there was one spot called Landore where the men (and their families if any) lived who operated the reverbatory furnaces and one heard considerable Welsh spoken there then . . .

When we (had) discharged the cargo we ballasted and went up the coast to Taltal to load saltpetre for orders. We had no steam so our cargo had to be taken in by hand winches (and) hove up the side from the lighters. The cargo gin (block) was guyed about halfway between the ship's side and the hatch coamings to give the necessary swing inboard to the sack of nitre(sic). The second mate ran the stage – holding the hook of the lowering fall in one hand and as the bag appeared above the (ship's side) rail the lowering fall was hooked on (as) the man at the lowering bitts took a turn. The winch men let go (the hoisting fall) and the second mate unhooked the winch rope, stood clear as the bag swung inboard and flung the hook (back down) into the lighter and the amazing celerity that was attained when every man was on his toes to beat the next ship at the game, was an eye-opener. The bag of nitre would be in the 'tween decks before it had time to swing across the hatch and the next bag would be on its way up the ship's side. Usually a 'stevedore' was employed to stow cargo. The sack was lowered onto a stack of sacks built shoulder high in the hatchway, up-ended by the man on the stack and the stevedore (then) carried it across his shoulders and stowed it in his place. The tiers of sacks were stowed from full width of the ship in the bottom (with) each (succeeding) tier (stowed) about 3 inches in from the next (tier) below, each sack resting on two in the tier beneath so that the whole was a compact mass. I never heard of a cargo of nitre shifting . . .

The cargo onboard, the last bag was 'shantied' up to the gin with a man astride it, and the usual sing-song that night with plenty of visitors from the other ships if there were prospects of reasonable refreshments in liquid form. And we prepared for sea (and our homeward voyage to Falmouth . . .)

From Falmouth we proceeded to Dunkirk to discharge and the crew were paid off. We loaded a cargo of straw and (were) towed to Swansea where the ship paid off the second mate, steward and me on Feb. 26th 1896 . . .

Postscript

The remark of the Yankee skipper on passing a bunch of copper ore ships around and about Cape Horn was that 'it looked like as if Swansea owned the whole darned Pacific Ocean'.

CAPTAIN JOHN JONES OF FERRYSIDE

In 1910 Captain John James of Ivy Cottage, Ferryside, published a privately printed booklet, priced 6d, entitled *From Ploughboy to Skipper*. It is a fascinating piece of local history, and I am able to quote from it through the kindness of his descendent, Mrs. Gwadlys Jones, of Killay, Swansea. During an adventurous career of over forty years at sea he seems to have sailed all over the world and it has been difficult, because of limitations of space to decide what to include.

John Jones was born at Blaengwenllan in Cardigan in May 1825, and first went to sea as a cabin boy at the age of fifteen on local coasting vessels. In 1841, he began a formal apprenticeship, spent mostly on vessels engaged in the north-east coal trade to the Mediterranean and Black Sea. In due course, having completed his apprenticeship and sailing as a mate for three years, he joined a new barque, the *Herald of Goole* as mate for a voyage to San Francisco, just in time to experience the consequences of the Californian Gold Rush of 1849. At this point Captain Jones can take up his story for himself:

> "About this time Captain Lawson was taken ill, and never made his appearance an deck again till we arrived at San Francisco. I was the means of saving his life by giving him a dose of Croton oil. This was the opinion of a doctor at San Francisco. Having arrived safely there we were nine weeks discharging cargo and ballasting, but this was all done by the crew, as the 'gold rush' was on, and we could not get any labourers under one dollar an hour. When coming into San Francisco the man at the helm, out of revenge for having to eat unfit beef a little time back, nearly ran into an American warship, and as I was on watch at the time on the bowsprit the captain of the warship fired at me, and the bullet passed within a few inches of my head. We then came to anchor for the night, during which we were boarded by 'the

wreckers,' who bound and gagged the watchman, and all the rest of the crew were taken ashore, whether they wished it or not, except two men and three boys. As the men were promised plenty of work at one dollar an hour, they did not much object. As we were lying at the wharf one Sunday, the captain (who was now quite well) and his wife, whilst walking the deck, observed his old crew that were kidnapped pass by. They stopped, and bawled out: 'How did he and his missus enjoy the stinking beef for food?' After having discharged and ballasted for Shanghai, we found it very difficult to get a crew. Captain Lawson made up his mind to start with the few whites he had with him, with a few blacks to pull the ropes. I did not like to risk my life with these blacks as part of the crew, so remained ashore. Captain Lawson sailed for Shanghai, paid off his crew, and shipped there for England a crew of Malays who engaged only to go to Malay. However, the Captain did not keep his word with them, and intended running them on to England, which they found out and mutinied, killing the captain and all his crew and scuttling the ship, the captain's wife going down with her. They took as much as they could carry of the ship's stores for sale, on some of which was the name of the ill-fated vessel. This raised suspicion, they were taken, put in prison, tried, and sentenced to death; eight in all. I then worked for some time at lifting a sunken ship in the harbour, which we got up at last, and for which I received pay at the rate of one dollar an hour. After this, as everybody had 'the gold fever,' including myself, I joined a party of seven, composed as follows: A doctor, a solicitor, two carpenters, a Scotchman, a preacher, and myself. As The Eldorado was on the Klamath River, about 200 miles from San Francisco, the next thing was how to get there. We accordingly held a council of war, and it was decided to buy an old ship lying at San Francisco, which we did for a mere song, and navigated her about 100 miles north. Of course, I was appointed captain. We landed in the Bay of Trinidad, and anchored off a small town called Trinidad, where we bought provisions, tools, donkeys, and mules. We then sold our ship for what it would fetch, and started for the Klamath River. On the way we slept one night in the hollow of a tree; redwood trees grow here to a height of 300 feet and 20 feet in diameter – and got to Klamath in about a week. We then set to work in earnest prospecting for gold, and found a good workable plot in a branch of the Klamath at Salmon Creek, and went on, with more or less good luck, until we heard that San Francisco had been almost burnt down to the ground (San Francisco was built all of wooden houses at that time). The two carpenters, the doctor, and solicitor left for San Francisco, as fabulous prices were given to any sort of labourer. The preacher left for Oregon, and only the Scotchman and myself were left at Klamath. We then worked at Orleans Bar for 12 months, when my partner left me to go home to his wife. I worked for a short time by myself, then I was joined by three Yankees, who helped me in bringing a supply of water to the Bar. It was conveyed for about a mile through ditches and troughs across a gully to where we were working.

I invented at this time a sluice to wash the gold from the sand, which even now goes by the name of 'Jones' Patent Long Tom'. We did well here, and made a lot of money. The Yankees left in 12 months, after having been paid out by me. Then I worked by myself for some time, till I took a partner to work, but my partner had no money. I worked with this man till the end, and sold my share to another man without any capital. He faithfully promised to pay me, but never did. I now left for the coast, en route for San Francisco, got to Trinidad, and caught a boat going to Portland, further north.

Whilst squatting on the Klamath River, we lived principally on what we shot, embracing stags, grizzly bears, buffaloes, game, and wild fowl, and caught all kinds of fish. We made our own bread from flour supplied us from the Stores. I have witnessed many times the way Indian children caught salmon. They used to watch the fish leaping at the falls, and with bows and arrows, shoot the fish, then go after them, and swim to shore with their catch. On many occasions we were greatly troubled with the Indians, which resulted often in sanguinary conflicts, and an Indian war-whoop is a thing never to be forgotten. We now left Portland for San Francisco, stayed there a week, and enjoyed myself very much. I should mention I carried the gold dust I had gained in a leather belt round my waist for safety. I then took ship for Panama, and when there paid one dollar for a hammock bed in a galvanised iron shanty, and was one of 400 (sic) occupants of the room.

I then started to cross the Isthmus of Panama by train (the first train that ran), which broke down half-way, and we were delayed eight hours. We had to cook our own food, and I paid some Indians 5s. for a small chicken and 1s. a glass for drinking water. I then started for Aspinwall, and from there to New York. Had a lot of fun at New York, where I stayed a week, and then sailed for London. We had a lot of trouble on this voyage with the sailors, who were a motley crew, and had never been to sea before. They were supplied to us by some New York 'bullies.' In about three weeks we arrived in London and had a very good passage on account of westerly winds. I remained six weeks in London sight-seeing and enjoying myself . . .

About the year 1860 I bought a share in the brig *Margaret* on the coasting and Baltic trade, outwards with coal, the last voyage being home to Hull with grain and deals. Then I went to L'Orient with coal, and back to Cardiff in ballast; shipped from Cardiff coal for Cadiz, and brought back a cargo of salt for a place near Revel. Off the Spanish coast a lot of fishermen hailed us, and for a bucketful of salt shovelled about half-a-ton of fish on to our deck. We pickled these, and did well in selling them when we got to Russia. Off Cape St. Vincent our ship sprung a leak, and we had to go and anchor at Lisbon Bay to repair pump boxes. We had them put right, bought six sheep at 4s. each from some Spaniards, and sailed away again. We killed the sheep, and put them in the hold with the salt. In a short time they were cured beautifully, and eaten with relish before we got to Russia. We had to call at Newhaven for a supply of provisions, and had here to clean the ship's bottom, as it was very foul with shell fish and seaweed. We got to Russia at last, after a long passage . . .

I made a few more voyages, coasting chiefly, and remained on shore to superintend the building of a new ship, in partnership with another man. This took about ten months, and the ship was christened *The Jane Emily*, after our two daughters . . ."

After a number of voyages round the Mediterranean, Captain Jones continues:

"We set sail for Glasgow and took a ship-load of coal and general cargo for Trinidad, in the West Indies, and I took my wife and family with me. We there loaded rum and sugar for Greenock. Before starting from Trinidad a clipper ship was about to start the same time, and the captain bet me a new hat he would arrive in Greenock before me. I won the bet by one tide, and would have done better still had not the rate of my chronometer gone wrong, so I had to be very careful. We left Greenock for Troon, where we took in a cargo of coal for Port Said, and thence to Alexandria in ballast. This took place when the Suez Canal was opened on November 16th, 1869. Unfortunately the broker made a mess of the freight and we lost money by the journey. We caught several sharks at Port Said, which swarmed with them, and saw one man who was bathing killed and eaten by them. We then sailed for Dublin with grain, and after discharging here went on to Swansea in ballast, where we loaded coal and coke for Algiers. Having discharged we then went in ballast to Tangiers and loaded Esparto grass for Cardiff. We also lost money by this voyage. My partner met me in Cardiff and was not at all satisfied. I explained to him it was no fault of mine as I had not been consulted but simply did what the

> broker had recommended, whereupon we mutually agreed to put the ship up for sale, so we tied 'the broom' to the mast and sold the ship in a short time in Cardiff Docks to a firm from Bridgewater. I then gave up the sea, after having been nearly forty years roughing it. This was in the year 1875."

CAPTAIN THOMAS GWYNNE EVANS

Captain Thomas Evans, born in Pembroke in 1843, was appointed Master of *Lothair*, a composite barque, 840 tons gross, built on the Thames in 1870 and owned by W. Bowen Jnr. of Llanelly, in 1884.

The *Lothair* had the reputation of being fast; it is said that in 1873 she made one of the quickest passages home from China, beating the *Cutty Sark* by three weeks.

The *Lothair* was engaged on voyages to China, Australia, British Columbia, Chile and Peru. In 1891, after discharging a cargo at Callao, Peru, she was sold locally.

(From an article by H. M. Tucker, *South Wales Daily Post*, 29.8.1953, and courtesy of Miss J. M. Evans of Burry Port).

CAPTAIN LEWIS, MASTER OF THE *ABERCARNE*

Captain Lewis, of Aberdare, has passed into history in an unfortunate manner, as the following extract from the *Cambrian*, 23rd August 1878, shows:

A Sea-Captain Mobbed at Swansea

> Yesterday between one and two o'clock the town was greatly disturbed by the tumultuous procession through Wind Street and other of the chief streets, of a mob composed of several hundred persons, men, women, and children, who were following the captain of a vessel which had just arrived in port, jeering and hooting, and threatening the infliction of Lynch Law as they went along. From the borders of the crowd, the astounded passers by in search of information, might learn that the 'captain' had 'starved his crew' or 'killed his crew', or 'ill-treated his crew'; that he was escaping from justice through the public streets in broad daylight, and that the mob were determined to do their duty by watching him if they did not Lynch him. In this spirit the crowd besieged him in his vessel, followed him from his ship at the docks to the office of Mr. John Dowle Jones, and thence through the streets until he came to the Castle Hotel, Castle

Square. The captain and his mate, who was in his company, turned into the Castle Hotel, and the crowd, thinking he would make his escape through the stable entrance, for the most part went round to Caer Street. While they were gone, the captain jumped into a cab and got off, and the mob, getting wind of his slipping away, divided itself and ran in different ways after him for some distance. Such a commotion has not been seen in Swansea for a considerable time past.

The facts of the case, as authoritatively stated, are these. The person who was thus followed was Captain Lewis, master of the barque *Abercarne*, lately owned by Messrs. Burgess & Shaddick, but now by Mr. John Dowle of Hamburg. On the voyage several members of the crew were stricken with scurvy, and one poor fellow – the carpenter, a Swansea man – died of it. It was alleged by the crew that the captain did not treat them properly when they were thus afflicted, and that if he had done his duty and applied the prescribed remedies, the carpenter would not have succumbed. The crew, or as many as were alive, were paid off at Falmouth, where a preliminary step was taken to bring the matter to an investigation . . . When the *Abercarne*, therefore, came into the port of Swansea, yesterday morning, she was literally besieged by the mob, who threatened the captain and prevented him coming ashore for some time, and followed him through the streets as we have said. The captain is, we believe, a native of Aberdare. A full inquiry into the affair will take place at Swansea in a few days under the direction of the Board of Trade.

CAPTAIN DAVID MORGAN

Captain David Morgan, along with Captain James Bevan of Horton, was perhaps Swansea's most celebrated Cape Horner, and we are fortunate to have a record of his life, written down in his old age, after forty years at sea, for which I am indebted to his descendent, Mrs. Phyllis Roberts, of Langland, Swansea.

He was born in 1836, and died at the age of 94 on the 4th September 1930 at Mumbles, where he lies buried in Oystermouth cemetery. I cannot do better than let Captain Morgan speak for himself:

"Briefly stated I went to sea in 1851. My father allowed me to go thinking that one voyage would satisfy all my inclinations towards a nautical life. But his hopes were in vain, for I was apprenticed during the following year 1852, at the age of 16.

During my apprenticeship my travels were confined to the coast of England, visiting London, Liverpool, Whitehaven, etc. My apprenticeship however was never completed. I ran away from my ship and took

Captain David Morgan and his family.
(Courtesy Phyllis Roberts, Langland)

berth as Ordinary Seaman on an American ship. She was bound for the Crimea calling at Trieste for hay. This was during the Crimean War. After this as Able Bodied Seaman I had several voyages, visiting New Orleans, Nova Scotia, New York, Rotterdam, Shields, Genoa, Leghorn thence to Boston. At Boston I entered the service of an American Revenue Cutter for 11 months. Our duties comprised an oversight of the Newfoundland Fisheries, and to guard against any contraband.

After leaving the cutter I visited the fresh water lakes of Canada, principally Ontario, carrying grain.

A longing for home now brought me back to the old Country. Having reached Gower, I remained there for some time and attended the nautical school at Swansea.

Another voyage or two, visiting the Mediterranean, calling at Galatz.

I made five voyages round the Cape, one as Able Bodied Seaman, two as Second Mate, and two as Mate. I obtained my Master's Certificate in 1864. The next 30 years were spent in long voyages round Cape Horn, altogether making 33 voyages, and crossing that dangerous cape 66 times. I carried copper, nitrates, guano, hides, etc.

To give you some idea of the faith we sea-faring men have in our calculations as to latitude and longitude, I have started from Valparaiso, sailed round the Cape, up the South Atlantic, through the North Atlantic and had not sighted land during the whole of that voyage until I reached Swansea Bay, a distance of between 7000 and 8000 miles. The calculations are made by means of the sextant for gauging the sun's altitude; the chronometer, a sea-watch which keeps wonderful time; sometimes the stars; a chart of course with which to localise the position of the ship when the latitude and longitude are found.

For twelve months at a time I never saw my wife and children, and my visits at home were often of very short durations.

American Sea Captains

Much has been written as to the cruelty practised by American Captains and their Officers. Many of you doubtless have read Frank Bullen's story *A Whalers Life*. That will give some ideas of the hardships endured by the sailors under their charge. I can vouch for the truth of much that Frank Bullen has written for I have seen some terrible blows given on board an American Ship when serving as a Ordinary Seaman. Such have been the wounds caused by the blows of these men that I have seen the blood run on the decks from the unfortunate who did not please them, But in all justice let me point out this fact in relation to this point. If you are thoroughly efficient they will not ill-treat you. It is when you do not know your work that the trouble begins.

I ran away from my apprenticeship at Liverpool. When I signed articles as Ordinary Seaman to a large American Ship the Officer was rather sceptical as to my abilities. I was only 18. 'What do you want', he said. 'When does this ship sail', I replied. 'What do you ship as?', 'Ordinary Seaman', I replied. 'At what wages?', '£2 a month'. 'O tip top wages!', 'Do you know your work?' I said that I did. Well said he 'Find me that top gallant buntling'. I had been used to all the parts of these American ships and in a moment was aloft and what he required was done. 'Now', said he (still not quite satisfied), 'we start on Monday but remember if you do not know your work you had better be somewhere else'. On coming out I passed two or three who had also agreed to join the ship. They asked me if I could steer the wheel. I thought they were making fun of me so I replied 'Of course not, but I can steer the ship'. Poor fellows they did not know what was in store for them. They had been shanghaied abroad (kidnapped) by means of drugs and their knowledge of seamanship was very imperfect. But they wished to get back so they signed on in this ship. On the following Monday we sailed down the Mersey. The second mate was in charge. He flourished a loaded cane and marched up and down the deck tapping everything near him in a most majestic fashion. The first order came like the roar of a bull, 'Two hands on each top, hook on the topsail sheets'. I started off and then called out 'Let go the topsail clew lines'. Alas the poor men could not find the ropes. The mate shouted to them 'You blackguards, let go the main topsail clew lines'. He helped them literally with his cane and his boots. It was pitiful to see them hurrying to and fro seeking some rope whose name they did not know, and belaboured by the cane, fists and sometimes boots of this hard hearted Officer. Life during that voyage was not worth living for them. Strange to say the title with which he usually addressed these poor men was 'Supernumeraries'. (This is a term which Methodists always regard with veneration, not so this American.) 'Come out you Supernumeraries' he would shout at all watches. Out they had to come at any or every watch. At first I came out too, but he pointing his finger to me said 'Now young Swansea, just go back, it is these supernumeraries I want'.

The Crimean War

During that terrible war I was before the mast in a ship carrying hay from Trieste to the troops in the Crimea. Calling at Constantinople I saw the ambulances laden with wounded on their way from the ships to the Hospitals. It was heart-rending to see the condition some of the poor soldiers. Several dying on the way. One must remember the imperfect condition of medical surgery, and the almost barbarous manner in which freshly amputated limbs were treated, such as cauterising with boiling oil, etc. The noble Florence Nightingale was then at Stamboul Hospital doing what she could to ameliorate the horrible state of things then existing. Strange to say whilst writing this I have just read in the *Daily News* of the Memorial Bust to Sir William Russell in St. Paul's. He was the first real war correspondent and drew attention to the sad way in which our soldiers were treated in regard to the Commissariat.

During the last great bombardment of Sebastopol I sat all night in the yards arms of our ship. I watched the terrible shells as they screamed through the air. The fearful noise as they burst was beyond description. At three o'clock in the morning the Russians blew up the powder magazine. Thunder was as nothing compared to this. Next morning the plain outside was dotted with tents. Our warships cleared for action ready to assault the Russians in the rear, but this was unnecessary; the end had come, the Russians had no more fight left in them.

I lost one of my crew very suddenly. It was blowing a gale of wind off Cape Horn. One of the top sails had loosened and was flopping about. One of the hands D... S... went aloft to fasten it. While he was about to fix it the wind blew it with great force against him, and knocked him off the top sail yards. He fell on the side of the ship striking the rail with his head, then straight overboard. Not a sign did we see of the poor fellow again. A mark of blood up on the rail showed the spot where he fell. I myself escaped death in a most remarkable way on one occasion. My rank at that time was mate. We were unloading coal at a port on the coast of Bolivia, called Loca Pelier. I was standing on a stage on the side of the ship directing operations, and moving the bags. Below was a lighter or barge, into which the coal was thrown. Outside lay another empty lighter. I was about to push a bundle of bags towards it when I lost my balance and fell head first down among the coal below. By a most providential coincidence I pitched on a bag of small coal. Had it been large coal my head must have smashed and my career ended. The sailors on deck were so frightened that they dared not look over. They expected to see my poor dead body. Imagine their surprise when I clambered over the side of the ship quite unhurt. The Captain was on the wharf and viewed this escapade, with feelings indescribable.

Earthquake

My experience directly of earthquakes has been somewhat limited, but my connection with the land of earthquakes has been of a very personal character. We all remember the record of that fearful earthquake (of last summer) which wrecked the town of Valparaiso. As I sat and read of the destruction of the various fine buildings I seemed to see them in my minds eye as vividly as I actually did so many times in the years gone by.

It was in May 1877, when a terrible earthquake took in Iquique Land and Areca. This was followed by a tremendous tidal wave which swept an American Man of War far up the country. My brother-in-law Captain R... J... arrived a day before this. He was on board his barque, *Bessie Jose*. The wave broke his ship from her moorings, snapped her cables, and tossed her about like a cockle shell. Fortunately that was all the damage done to her. Not so with others. One vessel was rolled over twisting her cable three times by this terrible wave, her masts snapping short and all on board drowned. A friend of mine Captain Trick of a Swansea vessel, the *Arctic*, was spending an evening on Board with a Bristol Captain and his wife. He thought it time to get back to his ship. The Bristol Captain begged him to stay just a little longer. 'Very well', said he, 'I will stay while your wife plays and sings 'God Save the Queen''. Only the Bristol Captain was saved. All the rest were drowned. His own ship was safe.

I was off the River La Plate at the time and the shock of this earthquake was felt on this river, right across the South American Continent.

Fires on Ship

With certain cargoes there is considerable risk with fire. This is particularly the case with coal. Sometimes a ship's cargo may be slowly burning quite unbeknown to the captain or crew. For instance a ship of which I was mate, on its voyage, after my leaving it, had a fire in its hold which had burnt itself out. The Officers of the ship were quite ignorant of it. How this was known is as follows:- When discharging the cargo one of the men disappeared from view among the coal. When they discovered him they found that he had fallen into a round oven like a room, the sides of which were smooth and burnt like coke.

When about to lade my ship at Swansea, I found that the coal in the wagons was very wet. There had been a heavy downfall of rain the night or two previously. Now damp coal becomes very combustible in the hold of a ship. I saw the owner about it. He said to me 'What can you do? It is your turn next to lade, and you will miss your stem' (as they call it) 'Well I shall say that I am not ready'. By and by, I was asked to bring my ship under the chutes but I declined on the plea of not being ready. 'Well you will miss your stem'. 'Alright', I said. However, I had not to wait very long and when I loaded I had the satisfaction of having dry coal. When I reached Valparaiso I was met by a Captain called Symons. He was the Head Captain in this Company. 'How is your cargo', says he. 'Alright', says I. 'Well', he said, 'I have been very anxious about you, several ships have been burning, one had to make for port on the way, and my cargo has been smouldering'.

I have known a ship so hot with the cargo burning that the heated plates have caused the steam to rise from her sides out of the sea. Officers and crew were afraid every moment that the rivets would burst. To crown all there was a large quantity of gunpowder on board, and they could not get at it by reason of the fire.

The Captain died from the worry of this, the nervous strain had been too much for him.

Nitrates which I often carried is under certain conditions very dangerous. Strange to say, if it is clean and free from substance such as sacking, paper, etc., it will not burn. But let any of these be about, a lighted match and in ten minutes the ship is ablaze.

Icebergs

During my many voyages round Cape Horn, out of 33 I have seen icebergs about thirteen times. On one occasion I have seen as many as thirty at once. This happened when we were rounding Staten Island and it seemed as if we were in the midst of a bristling mass of ice giants. Again I have gone so far South as to reach a large icefield, and had to return out of its dangerous presence. Once on my return journey from Chili I had the pleasure of encountering an iceberg very far north, being in fact about 34°S. It was a beautiful object and glittered in the sun like gems. The temperature of the surrounding air was very much affected by it and the thermometer registered much lower than usual. The part of the iceberg under water is very much more in proportion than that above, being about two-thirds to four-fifths. This under portion is continually melting as the iceberg moves in the warmer waters. As this process continues the iceberg is in danger of losing its equilibrium. When we approached this large iceberg I imagine my ship's presence caused the required disturbance, for just then it turned over with a thundering rending crash. There was a cracking of ice like the noise of a cannon. Pieces flew about with terrific force. Fortunately we were out of harm's way, and so enjoyed one of the most beautiful and awe-inspiring sights that falls to the lot of a sailor.

Struck by a Sea

When I was the other side of Cape Horn we encountered very heavy seas. Very few landsmen have any idea of the force encountered by any object when meeting a strong wave. I was just going below when a fearful sea struck one side of the companion, breaking the glass and throwing it against my chest. I did not give this an particular attention at the time but when I was about to lie down for a few hours I felt that my chest was wet. On examination I found that it was moist with blood. A piece of glass from the companion way had penetrated my clothes, and cut my flesh. I had on at the time a thick oil-skin coat, an overcoat, an ordinary jacket, a waistcoat, a jersey and vest. Yet the piece of glass had cut through all these and inflicted a rather deep flesh wound. It must be borne in mind that an oil skin as worn by sailors is tough as leather and much harder. The overcoat was also of

> thick felt material such reefers are made of. Yet such was the impact caused by the strength of the sea that all these articles of clothing were penetrated."

Captain Morgan was in command of the *Serena*, a composite iron barque of 493 tons, built at Sunderland in 1866 and owned by H. J. Madge, when he had to abandon his ship off Cape Horn, following storm damage. It was lost in latitude 5834S, longitude 6950W on the 3rd August 1893. Now let him take up his story again:

> "We were outward bound round that stormy head-land, Cape Horn. Already we had experienced a lot of dirty weather. No sooner had we passed Staten Island when a terrible sea struck the ship forward. The force of this blow loosened the wood-ends from the stem and the sea began pouring in. All hands that could be spared were immediately put to the pumps.
>
> It was bitterly cold, freezing, snowing and blowing a biting gale. The mate was aloft helping to set a bit of canvas to enable us to keep her head to the wind. He had become so benumbed that he could not move without danger to himself. With great difficulty I climbed up the rigging and brought him down under my arm. We had a fortnight of this weather, beating to and from near the Cape on account of stress of storm and gale. The waves were as high as a house, and the impact on the rudder caused its gearing every moment to be in danger. I was at the wheel on the poop. Three times the sea caught me like a cork and threw me over the rail but I clung on to that rail with the grip of death, knowing that if I let go I should be swept into eternity. Just then being insufficiently clad the mate said to me, 'You had better go below Captain and change, I'll hold on a bit'. I never saw him again. I had just got to the companion ladder when a huge sea enveloped the ship on both sides. This is most unusual, the sea generally striking the vessel one side or the other, or fore and aft. In this case the waves seemed to envelope or fold over the ship. Fortunately she was going at a great speed, otherwise she would undoubtedly have been swamped. As it was the mass of water on deck pulled everything back with it. As I descended the companion way, unaware of the approach of this sea the sides were smashed in on top of me. Two sailors were also thrown down into the cabin. When I returned to deck, I found my two Officers, the first and second mate, were gone; two men were also lost. I went for the wheel but alas that was gone and the rudder also. There we were at the mercy of the winds and waves. The sail was blown to ribbons. The men had to keep at the

pumps to prevent her from sinking. I had given up all hope. We had drifted down very far south out of the track of other vessels. Providentially however, another vessel had been driven farther south than we. We drifted helpless from the Sunday until the Friday when we sighted her. She took us on board and in 23 days we reached Valparaiso. There I had to undergo an examination by the Board of Trade as to the loss of my ship. I was given a Certificate of Exemption from all blame. From Valparaiso I reached home by mail steamer. Unfortunately my wife and children had heard that my ship was lost and that two Officers and two sailors were drowned. They were in great suspense until another welcome telegram assured them that all was well. I carried the marks of this storm for many a day. The Strain of holding on to that rail turned my nails and fingers black and they remained so for many weeks.

Shipwrecks (Rescue)

I have been fortunate enough during my Captaincy to rescue four ships crews, though one of them could scarcely be described as a rescue.

During my first voyage in command, on my outward course we came across a vessel from Bristol laden with coal. She was sinking fast. She had on board a crew of about fourteen or fifteen. That very afternoon I placed this crew on board another vessel homeward bound, thus saving them an unnecessarily long voyage.

One night my ship ran down a Dutch Brigantine from the River La Plate. She was clearly in the fault. Out of her crew of nine I managed to save only two. Strange to relate 17 days after I picked up the crew from a Glasgow ship on her way to Peru. She had a large quantity of machinery on board. With the nine men from her and the two already on board I proceeded on my voyage. I left them all on the Falkland Islands.

Last but not least I had the pleasure of rescuing the crew of a Norwegian barque. She carried a cargo of teak, from Demerara. Now teak is a very hard wood and will float. This cargo had shifted and was knocking the ship's sides about until she was in a very leaky condition. For several days all hands had been at the pumps. We saw her signals of distress and made for her at once. She was situated between the Azores and Cape Verde Island. The crew of nine were safely brought on board. I left them with a Pilot Boat who took them to Queenstown. The King of Norway and Sweden sent me a Silver Medal for doing this bit of what any Englishman would regard as a pleasant duty. It was presented to me by the Mayor of Swansea in the Town Hall.

Conclusion
And now my sea-faring life is over, yet I live in the past, recalling vividly the many scenes and incidents of bye-gone times. I love, again and again, to spin those old sea-yarns to my son-in-law who is just like a school-boy in his thirst for sea stories."

ANDREW NELSON

Andrew Nelson was born in Sweden and christened Andre Neilsen. He married a Swansea lady, Harriet, and changed his name to Andrew Nelson. He was an interesting link with the *Caswell* (see page 159) for he sailed on that vessel on the first voyage after the mutiny, together with Dick Sullivan (see page 153). After retiring from the sea, he kept a shop in Morriston, where he died at the age of 82.

HENRY JOSEPH NICHOLLS

Henry Nicholls is as good an example I have come across of the best type of Swansea seaman who never aspired to command but was supremely master of his own craft as an able seaman, sailmaker, and finally boatswain.

Through the kindness of his descendent, Mrs. Elvira Bevan, of Swansea, I have copies of his discharge papers from a whole series of the most celebrated of Swansea copper barques such as the *Ottawa*, *Tacna*, *Glanafron*, *Llewellyn*, *Hinda*, *Rossyn* and *Uncas*, together with his picture, which appears on page 128.

His discharge certificate from the *Hinda* is particularly interesting because it shows that he signed on in the dual capacity of sailmaker and able seaman.

He was born in Jersey in 1854 and died in Swansea in 1923 at the age of 69 after serving for many years as a Swansea boatman, and later, dock policeman. His obituary describes him as 'a man of superior education, and a first class speaker, and knew the origin of dock troubles better than most men.'

He was apparently a retiring and modest individual. Once he was shipwrecked in the southern seas, but rarely spoke of it.

CAPTAIN J. E. OWEN

Captain John Edward Owen was born in 1877. On his Master's certificate, dated 28th December 1903, his address was given as 56, Gorse Lane, Swansea. In March 1936, after he had been many years in command, he was living at 22 Glanbrydan Avenue, Swansea. His first voyage to sea was on the Swansea ship *Kildonian*, a 650 ton iron vessel later cut down to a barque, built at Glasgow by Stephen & Son, with a cargo of coal for Buenos Aires.

Fortunately, in 1956 he was persuaded to write down his reminiscences, and in them, we have, stewed in their own juice, the authentic voice of a typical Swansea Cape Horner, a man who went to sea as a ship's boy and rose to command.

On the 1st June 1898, he signed as A.B. on the Swansea barque *Glanariff*, 468 tons, a flush-decked iron vessel built in Dumbarton in 1869 by Mcmillan & Son, and owned by Morgan Tutton of Swansea.

I give below, by courtesy of the Swansea Maritime Museum, his account of the voyage:

> "On June 1st 1898 I signed on as A.B. on board the barque *Glenariff* of Swansea, owned by Mr. Morgan Tutton. She was a nice little vessel of 450 tons, 161 feet long, 28 feet beam and she had a flush deck, i.e. no poop. We sailed a few days later with a cargo of coal for Punta Arenas, a small port in the Straits of Magellan. Our crew consisted of Capt. 2 mates, cook, 6 A.B.s, 1 ordinary and 1 boy, 12 hands all told. Our accommodation was down the fore peak, were (sic) the ships sides were bare iron white whitewashed, and a double row of bunks on each side. There were lots of rats on board and they often ran over our faces when we were asleep, but sailors don't mind rats.
>
> The forecastle was lighted by 4 port holes and a tin lamp, and on the whole wasn't a bad place, we even had a bench to sit on. Coal for the Galley had to be got up from the lower peak through our forecastle but this happened only once a week.
>
> Perhaps you would be interested to know what we had to eat in those days. Well by Act of Parliament the owners were obliged to supply us with a minimum quantity of food. There was no act stopping them giving more if they so wished. Very few of them did.
>
> *Monday.* Breakfast: porridge made of rough oatmeal sweetened with molasses that tasted like tar. This dish was mostly thrown overboard. Dinner: pea soup and salt pork with potatoes if any and there seldom were any after 6 weeks at sea. Tea: just a pot of tea and what you saved from your dinner. Nothing else supplied by the ship.
>
> *Tuesday.* Breakfast: Wet Hash a wonderful mixture of water, potatoes and fat salt meat. Only a sailor had a stumoch (sic) strong enough to keep it down. Dinner: Bean soup salt beef, 1 lb. fresh bread. No afters. Tea: as before.
>
> *Wednesday.* Same as Monday.
>
> *Thursday.* Same as Tuesday.
>
> *Friday.* Same as Monday with Salt fish for breakfast instead of porrage.
>
> *Saturday.* As Thursday with Rice and Curry for breakfast instead of wet hash. Dinner: Salt beef Bean soup. Rice pudding made with

rice left over from breakfast mixed with a few currants smothered over with fat from the salt meat and baked in the oven (not too bad) and 1lb fresh bread. Sunday was the star turn of the week. We had dry hash for breakfast made with potatoes if any, if not then made with biscuits soaked in water (not so good) . Dinner was Tinned soup, Tinned boiled meat, potatoes if any and what was known as Plum Duff nearly always eatable. Tea as before.

Lime juice to keep away scurvy was served every day. Our weekly stores were ½lb. butter, 1lb. marmalade, 12 to 14 oz. brown sugar, no bacon, cheese, jam, pickles, eggs, Tinned fish, tinned meat, milk or fruit. Plenty of ships biscuits. Water was carefully measured out to us, 3 quarts per man per day, and the cook took most of that for cooking purposes. We could only wash ourselves about twice a week then 2 of us used the same water. Clothes were washed in water saved when it rained. Food such as meat, and potatoes were cooked in sea water, bread was also made with sea water.

Leaving home summer time we had a good run down to the entrance of the Straits of Magellan which we made after about 80 days. We arrived there getting on for dusk and the Capt decided to stand off the land until the morning. During the night the wind came off the land and we didn't make it again for a week.

The Straits of Magellan is not an ideal place for a sailing ship. Anchorages are very few and the wind uncertain. We went through the 1st narrows in a dead calm carried along by the very strong tide and we were turning around and around as we went along.

There was no watch below now everybody had to keep handy and take what rest we could and when we could. After passing through the narrows the wind came favourable and we arrived at our destination safely.

The cargo was discharged into a store ship laying at Anchor. No shore labour was employed but we had the use of the hulks steam winches which were a great help to us.

After the cargo was out and ballast taken in we filled up with fresh water by towing our lifeboat with the jolly boat to a spring on shore were (sic) we filled the lifeboat with water in bulk towed it off to the ship and baled it into our water tank with buckets.

We were now ready to proceed to Iquique where we were chartered to load a cargo of Salt Petre. The Capt decided to sail right through the Straits and waited for a fair wind to do so. But our luck was out and after waiting a week he gave it up and we went out the way we came in and around Cape Horn.

Well you have all heard of the heave (sic) seas and gales of wind off the Horn. It is mostly so but there are seasons of Easterly winds which are fair winds. And we were lucky enough to get around with all sail set and soon got up into fine weather latitudes.

A week before arriving at Iquique we closed with the big 5 masted French Barque the *La France* and kept company with her for several days, the wind being light and right aft but once the wind shifted more on the side when all her sails would take wind she just walked away from us and she arrived in Iquique 2 days before us.

The *La France* had a large crew and steam winches and could work all her holds at the same time she loaded and sailed before we got our ballast out. There were 44 sailing ships and 4 steamers laying at anchor loading when we arrived, and it is the custom when a ship finishes loading for one of the crew to sit on the last bag and be hoisted up with it right up to the block so that other ships can see him. He then cheered his own ship and the nearest ships to the right and left of him.

And after tea the custom was for the Capt to send a bottle of Rum forward and under its cheerful influence all hands would get on to the forecastle head and cheer every ship in the harbour and they had to give an answering cheer, if they didn't other ships passed uncomplimentary remarks about her. If the finishing ship didn't start the cheering it was understood that no bottle was forthcoming from the Capt and he got his fortune told by the ships near enough to tell him.

The cargo comes alongside in small lighters sculled off by 2, 3 or 4 men according to the size of the lighter. The oars are not pulled on the side but sculled over the stem, the men keeping time with each other, and the bags of Salt Petre are hove on board, one at a time, by means of a hand winch, and is stowed prymid (sic) fashion the ships sides being clear of cargo except the bottom tier of bags which is stowed all over the bottom every other tier being ½ bag less than the one underneath it. When loading Salt Petre barrels of water were kept handy in case of fire.

After the cargo was loaded we were allowed a liberty day and it happened mostly from Saturday night to Sunday night. This was our one and only chance to have a run ashore during the voyage.

The ship never being alongside a quay from the time she left Swansea until we got to the discharging port. Only half the crew were allowed ashore at a time and when they returned on board they generally had a supply of the native poison to carry on the celebrations for a little longer.

In Iquique there was a tug boat to assist vessels in or out but in some of the West Coast Ports where there is no tug and lots of calms vessels were helped in and out by other vessels' boats.

After loading was completed we sailed for Queenstown for orders. We had light winds leaving, and a couple of days later were becalmed for a couple of weeks. It is not a pleasant experience to be becalmed with a cargo of Salt Petre. There is generally enough of a swell to keep the vessel rolling and the way a heavy loaded vessel can roll has to be experienced to be believed, sails flapping, pots and pans had to be lashed on the stove creaking and groaning of the masts and spars, and to try and eat pea soup would be a star turn in the Empire if it could be reproduced there.

We would sit on the forecastle floor back to back with our feet up against something solid and balance our dish in our hands and were real artists at the job. One had to be alert to get a meal.

As usual when homeward bound, when in the N E trade wind it is the custom on some ships when painting ship, for the watch due to go below for the afternoon, to have to stay on deck and work. This was a most unpopular event and caused a lot of grumbling.

We had to do a couple of weeks of it and when the wind was light and vessel fairly steady we put stages over the side and painted the ship outside. In due course, we arrived at Queenstown for our orders, and some food, we were down pretty low by now. As usual the Bumboat came aft and did a roaring trade especial the boat with eatables. They sold everything you could wish for except drink. The Tailor and bootmaker also did well. And as the Capt had to settle the bills and deduct the amount from our pay he also had his usual commission and could afford to smoke a nice fat cigar. After a few days we received orders for Hull and while towing out of harbour it was my job to hook the cat block to the anchor and thus hoist the anchor up to the cat head. To do this I had to go over the bow and while on the anchor the tow rope slacked up and knocked me overboard.

My first thought was 'don't get under the ship's bottom' and I struggled clear. On deck they were shouting 'throw him a rope' and out of the hundreds of ropes on a sailing ship I didn't get one, but the men on the tug had seen what happened and stopped the tug. Our boat was hanging over the side from the davits. They lowered it and in a few minutes I was back on board again. The Capt was really glad to see me come aboard (it would have been a lot of bother for him if I hadn't). In fact he was so glad that he gave me a real good tot of rum and sent me forward to change my clothes. It all happened in about 20 minutes from the time I fell over until I had changed my clothes and stood at the wheel feeling on top of the world after my swim and drink of Rum, and besides being at the wheel I got out of all the hard work of hoisting yards and making sail.

We arrived at Hull a week later and when I landed on the wharf to make the moorings fast a man standing there said to me, 'What part of Swansea do you live?'"

(*The above is quoted courtesy of Swansea Maritime Museum*).

CHARLES PRITCHARD, SAIL-MAKER AND SEAMAN

Charles Pritchard was born in 1851, and is particularly interesting because he served a seven-year apprenticeship as a sail-maker with the Swansea firm T. P. Richards & Co., whom he joined at the age of sixteen. One of his certificates of discharge, for which I am indebted to his descendant Mr. Charles Pritchard of Merthyr Tydfil, together with his photograph on page 152, shows him as joining the Liverpool ship *William* at Buenos Aires in August 1870 as ordinary seaman. He must have therefore left some previous vessel in that port, probably due either to sickness or accident.

He spent more than thirty years in the employ of T. P. Richards, and most of his voyages seemed to have been on the Cape Horn run to Chile, Peru, Oregon and San Francisco. In 1878 he joined the famous Swansea iron barque *Arctic*, 616 tons, and built at Glasgow in 1855 by David and Wm. Henderson, and continued to sail on her until July 1882, when he left her at Plymouth.

He died in Swansea on Christmas Eve, 1925, and the large number of mourners testified to the respect in which he was held. At the time of his death he was caretaker at Gloucester House, Swansea.

THE REES FAMILY OF SWANSEA

There were many families of seafarers on Swansea during the 19th century, and one such was the Rees family. Captain John Rees, born in the hamlet of St. Thomas in 1792, commanded a well-known local schooner, *Wave*, in the local coasting trade. His sons, Captains David, Samuel, Thomas and James Rees, sailed in the larger Swansea copper barques which proliferated during the copper boom of the 1830s to 90s. Captain James Rees, the grandfather of Mr. Colin Rees of Mumbles, to whom I am indebted for this account, married a Miss Hinckley, daughter of James Hinckley, the manager of the famous Cambrian Pottery from 1824 to 1851. Her grandfather was Captain William Buckler, born 1780, master of the Clovelly vessel *Hope*. Her brothers, James Henry, William Buckler and Charles Hinckley, were also captains, the latter in later life commanding the steam-assisted vessels which appeared towards the end of the century.

Captain James Rees and his brothers and brothers-in-law sailed in some 75 vessels, often in ones which their brothers had been in previously, involving many passages round Cape Horn to the west coast of South America, and also to South Africa, Cuba and Newfoundland. His picture appears on page 152. Sadly, Thomas Rees was lost with all hands on a voyage round Cape Horn in 1876 in the *Herradura*.

Charles Pritchard.
Seaman and sail-maker.
(Courtesy Charles Pritchard, Merthyr Tydfil)

Captain James Rees and his wife Annie Eliza Rees (née Hinckley).
(Courtesy Colin Rees, Mumbles)

Dick Sullivan, boatswain.
(Courtesy Audrey Roberts, Killay)

Captain William Tanner. CapeHorner and Swansea tugmaster. See business card, page 69.
(Courtesy John Tanner, Swansea)

Thomas Wheatland, carpenter.
(Courtesy John Wheatland, Kidwelly)

James Hinckley, Master of the *Maggie*, owned by Gillespie of London, died in Tobago, while Charles Hinckley died on entering harbour at St. Johns, Newfoundland. His younger brother John was a sixteen-year-old apprentice aboard the barque *Pearl* when he died of yellow fever in Cuba in 1855. So the Reeses, like other Swansea seafaring families, paid a heavy toll for their contribution to the history of Swansea in the days of sail.

THOMAS REES OF LLANGENNITH, GOWER

Through the courtesy of Mr. Dale Thomas of Monmouth I am able to give below some details of the life of his great-uncle, Thomas Rees.

Born in 1873 Thomas Rees went to sea as an apprentice in the iron barque *Andaman*, 919 tons, built at Port Glasgow in 1865 and owned by Simon Goldberg of Swansea. After completing his apprenticeship and qualifying as a third mate he signed on the *Afon Cefni*, a large four-masted steel barque of about 2,100 tons which sailed from Swansea, 5th January 1894, with general cargo and coal. She was owned in Anglesey by R. Hughes & Co. of Menai Bridge, and their record book contains the melancholy story, 'Lost, ship never heard of again'.

In a letter written to his father from the *Andaman* dated 7th September 1890, Thomas writes '. . . hoping all the people are well and enjoying themselves before the fire on a cold winter night when we are up here by night as well as by day every other four hours we are on deck at night. Xmas won't be long now but it will be over before you get this. I wish I was at home instead of out here out . . . Had I known as much as I do now no *Andaman* would have got me to sign working here two years in January have not got a farthing yet the men say if they were me I would run away. It is very hard to come out here for nothing . . .'

But Thomas stuck it out in the end, and got his certificate – only to lose his life on the *Afon Cefni* four years later. The family grave in Llangennith churchyard, still visible today commemorates his brief life.

WILLIAM RICHARD SULLIVAN

Dick Sullivan was born in Hafod, Swansea, in 1872, the son of a Cape Horner, James Sullivan who had sailed as a Mate on the *San José*, a 393-ton wooden barque built at Sunderland in 1861 by Pickersgill & Miller, owned by Richardson Bros.

I am indebted to his descendent, Mrs. Audrey Roberts, of Killay, Swansea, for the following information about his life: He died at the age of 94 and until shortly before his death he walked every day from Odo Street in Hafod to Swansea, in spite of having only one lung as a result of being gassed while serving in the Royal Navy during World War I.

He was one of four brothers and a sister, and one of his brothers also went to sea, because family tradition has it that Dick Sullivan was at one time going in to San Francisco on a ship while his brother was on board one that was due to sail. His brother asked permission of his captain for time off to visit Dick, but this was refused, so he deserted his ship and met up with Dick anyway.

Dick's youngest brother Harry owned Sullivan's Canteens on the Swansea docks, which prided themselves on having kept open throughout World War II in spite of the severe bombing which destroyed most of the town centre.

Dick was a distinct character; Chairman of Swansea Football Club, and a leading light in the Swansea Ship Lover's Society. In 1956 he gave an interview to the Swansea *Evening Post*, which published on the 19th, 21st and 22nd of November of that year, and the following extracts from it give a flavour of his life.

> "'Pound and a pint'. The phrase used by Dick Sullivan intrigued me during his recital of a memory of his old Cape Horn days. I repeated it as a query, and he came back at once with the reiteration:

'Aye, man, pound and a pint, pound of meat and a pint of water as the day's ration, and some bread if the flour had lasted out and the weather was easy enough for the cook to bake it.'

"What we didn't eat for dinner we kept for breakfast," he added. "There was not much variety in the grub, meat, generally salt pork, hard biscuits with occasionally some bread, and pea soup when it could be cooked. But there were days on end when we had no hot meal at all because the galley fire was out as a result of the seas."

It comes as a shock to learn that here is an almost direct link with one of most striking true stories of the sea, the mutiny on the *Caswell*, one of the goriest tales of murder on the high seas, for Dick Sullivan sailed in that ship in 1886, only a few years after the gruesome incidents which startled the country in those far off days of the last century.

'I stowed away in the *Tacna*, a barque belonging Mr. Hennessy, the jeweller in Wind Street. Unfortunately before we passed Lundy I was discovered and, broken-hearted, I was returned to Swansea in the tug-boat.

All my pals were apprentices in ships, but my turn was to come. When I was 15 my father successfully applied to find me a billet on the barque Caswell, under the command of Captain Henry Lewis, and thus I started my career as a Cape Horner.

In the years that followed I served in several ships: the *Kappa*, *Foxhound*, *Thunderbolt*, *Maxima*, *Delta*, *Beta*, *L'Esperance*, *Larnacha* – a full rigged ship – and the *Gunford*, a four-master.'

The accommodation for the crew was very scanty, a ship ladder giving access to the forecastle, and the crew sleeping in bunks on either side.

There was no stove, and the only light was one improvised from a tin of grease with the threads of canvas for wick.

The small galley or cook-house had room for one man only, the cook holding the dual position of cook and steward. Long voyages, the absence of proper storage – the 'fridge' had not then been invented – meant that the food was coarse with little variety. Fresh meat soon ran out after leaving port and there was a return to the inevitable salt beef and pork. Ships' biscuits became riddled with weevils and by the time the Cape was rounded rations were meagre, and often there was no cooked food for weeks, the cook being unable to use his galley through bad weather.

The captain's quarters were also very cramped. Very few of the ships carried a second officer, but instead, a boatswain, who usually slept in the half deck. The barques usually carried five or six apprentices who also occupied the half deck, like the boatswain and the carpenter.

The half deck was adjoining the cabin with an entrance to it and at the break of the poop. The apprentices usually started at the age of 15 or 16, and it was customary for their parents to pay a premium of £40.

'I remember starting off on a voyage in one of Burgess's ships one Monday afternoon. We got aboard at the locks, some of the crew, as usual being very groggy. It was soon dark, and the wind veering round to the east, the order came for all sail to be set, as captain wanted to get south as quickly as possible. Fore and main topgallants were ordered, and I saw some of my mates go aloft the foremast. We on the main set our topgallants, but the foremast topgallants had not had the gaskets taken off, and I went aloft to see what was the trouble.

To my surprise I found that the man already aloft, a Russian Finn, had the horrors and was holding on to the rigging screaming. It was a difficult position up there on the tree, trying to cope with a nerve-stricken foreigner the while clinging on to prevent myself falling to the deck. I whipped out my knife and cut the gasket from the yard, lashing him to the rigging until I could get help.

The wind by now became stronger, and it was no good trying to shout as no one on deck could possibly hear. Therefore, I went down the stay hand over hand and let the captain know what was wrong. More men were soon up the rigging, and with the usual gantline we lowered him to the deck.'

Another experience, but this time at the end of the outward voyage, is recalled by Mr. Sullivan:

'I sailed from Swansea in the four-masted barque *Gunford* of Greenock, master Captain Weir, with a cargo of coal. We arrived at San Francisco after 110 days. On April 8, 1896. At about eight bells in the afternoon watch, and anchored about a quarter of a mile from the *Blairmore*, which was light, and awaiting loading.

The following morning we were unbending sails when we suddenly had a series of very heavy squalls. The *Blairmore* was swinging round on her cable, when suddenly she was struck by a terrific squall, which caught her squarely, and over she went.

Ours was the only ship in the bay with her, close to the long wharf and the big iron works. The captain of the *Blairmore* had his wife with him on that voyage and they had left the ship for the shore about half an hour before the ship capsized.

The crew were in the hold, chipping and painting, and it was very pathetic to hear them knocking with their hammers at the hull of the ship.

A squad of men came off from the iron works and a hole was made, but the air came out quickly and down the ship went with the men inside her.'

A notable feature about the ships was that their owners included quite a large number of local tradesmen, and among those who had an interest in them were Gale, the provision merchant of High Street; Fulton, the wine and spirit merchant; William Jenkins the butcher; Down, the metal merchant and some ship-chandlers.

Some of the ships were 'single-owned'. They were generally barque rigged, that is to say, square rigged on the fore and main mast with a spanker and topsail on the mizzen mast, and forward on the jib-boom, an outer jib, a flying jib and an inner jib, a fore topmast staysail.

When the ship was ready for sea, she was usually brought into the basin by the riggers, most of whom were themselves old Cape Horners but too old to continue at sea.

At the lock gate there would be the familiar cry of the Board of Trade Officer who would call 'all aboard' for the crew to take over. It might be that there were one or two short and a last-minute search of neighbouring hostelries, of which there was then a surprising number in the vicinity of the Strand, would be made.

Failing result, substitutes would be put aboard. The answer was given to the tug-boat, the pilot would order 'Let go aft' and the ship was away.

For a long period in the busiest days of the Cape Horners a famous pilot was Capt. John Rosser, and his foreman, J. Wilkie. The cheering of hundreds of relations and friends was the last link with the shore that those aboard would have for months.

Often, through the long delay waiting for a fair wind, there would be six or seven ships outward bound together. Off the Mumbles one of the small sailing craft which then served as pilot boats would come alongside to take off the pilot."

Dick Sullivan was not very long in steam and a somewhat pathetic story arises from the gradual decay of the old sailing ship trade. With his experience as a bo'sun on the sailing craft he was offered a similar post in a steamer.

The Captain suggested that he should vet the crew and when all were signed, mentioned that there was one other man to be signed on.

It was for a personal reason, declared the Captain. The man was a former friend who had fallen on evil days. 'Go to the bottom of Wind Street and ask for so-and-so', said the Captain.

'Sign him up. He is old and will not be as good as a younger man, but we can give him a few brasses to clean and other little jobs to give him the knowledge he is doing something.'

While waiting under the railway bridge, Dick Sullivan was spoken to by a man he was very glad to see, the captain of a ship in which he had rounded the Horn. They exchanged memories until the former shipmaster said, 'Well, I must not keep you, bo'sun. I am expecting to see a man who has offered me a job on the . . .'

'It was a strange position for me', says Mr. Sullivan. 'I said, "Why. You must be the man I am looking for", and it was the case. I signed on my old captain to serve before the mast'."

CAPTAIN JAMES THOMAS

Captain James Thomas was born in 1832 and obtained his Masters certificate at Liverpool in 1858. His career is interesting because eight years later he became master of the newly built *Powhattan*, one of Swansea's famous copper ore vessels and commanded her for most of her subsequent career of seventeen years under the British flag. Between 1866 and 1876 she traded largely to the west coast of South America. After that she was engaged in voyages to North America, Greenland, Iceland, South Africa, West Indies and the east coast of South America – Brazil and the River Plate.

The *Powhattan* was built in Swansea by the Meager brothers in 1866. She was 400 tons, and rigged as a barque, with what was described as a 'full male figurehead'. In 1883 she was cut down to barquentine and in May 1890 she was sold to 'two Spaniards'.

His great great grandson, Wing Commander Graham Hingston-Jones, O.B.E., of Tavistock, is actively researching further into the family history.

WILLIAM THOMAS, SHIP'S CARPENTER

Mrs. Barbara Davies of Langland has kindly provided me with the following information about her ancestor, William Thomas, who sailed as a carpenter on many of the famous Swansea copper barques. He must have had somewhat of a charmed life, for he was on the *Hecla*, which brought yellow fever into Swansea in September 1865, and on the ill fated *Serena* (see page 144). He was ship-wrecked on the *Devonport*, rounded Cape Horn twenty times and made the voyage to yellow fever-ridden Cuba thirteen times. After leaving the sea, he was for many years a shipwright in Swansea. He was the second son of Captain James Thomas of Swansea, and spent his last years at the Tunnel Hotel in Swansea, owned by his son, where he died in 1908 aged 75.

THOMAS WHEATLAND, SHIP'S CARPENTER

Mr John Wheatland of Kidwelly has told me of his grandfather, Thomas Wheatland, for which I am indebted to him.

Thomas Wheatland was born in Hove in Sussex in 1870 and went to sea at the age of 14, later becoming a ship's carpenter. Interestingly, a large part of his seagoing career was spent on ships running from Swansea or Liverpool to San Francisco. In 1894 he deserted in that port, and spent the next two years largely working on farms in the area. Mr. Wheatland recalls many conversations with his grandfather, who had strong views and was fond of disabusing what he felt were popular misconceptions. One of these was that all Californians in the 1890's walked round carrying revolvers; the only one he saw was carried by the local Sheriff. (In contrast, carrying guns in South America was general.)

Before signing on a ship's crew would always enquire about the master, knowing that their lives would depend on his knowledge and skill. Germans were unpopular because they had a reputation for carrying too much sail in an effort to outdo all others, and the most popular, according to Thomas Wheatland, were Norwegians.

He was shipwrecked twice; once totally for which we have no details, and again in Patagonia, where the vessel spent ten months being repaired. Sadly, the names of Thomas Wheatland's ships have not survived; all we have are his grandson's recollections of conversations with the old man. According to him, the term 'Skyscraper', first used in America, originated from the look-out point on a ship ABOVE the crow's nest. He was very suspicious of Chinese, warning his grandson 'never turn your back on them; they always carry knives'. Finally, he had an extraordinary story that on one of his ships, the *Dumfriesshire*, the master, Captain Lewis, died on the voyage from San Francisco to Swansea and, as a mark of respect, the crew tied up his body near to the wheel as the vessel entered Swansea. It is an odd tale, that I have tried, without success, to check up upon, and which I include because it does not sound like an invention.

Thomas Wheatland died in April 1940, and lies buried at Holy Trinity Churchyard, Pwll, Llanelli, where his tombstone is still clearly legible.

10.

Voyages and Incidents

THE LIST OF DISASTERS, strandings, shipwrecks, burnings, accidents, murders and cases where vessels just left port and were never heard of again is long enough to fill a book in itself. Here I have tried to give some of the most representative and interesting.

THE MUTINY ON THE *CASWELL*

On May 19th, 1876, the *Cambrian* newspaper reported the arrival of the Swansea barque *Caswell* at Queenstown, Ireland, with a cargo of nitrates from Antofagasta. There had been a mutiny on board, in which the captain, first and second mates, and a coloured steward, had been murdered by five members of the crew, three Greeks and two Maltese.

The *Caswell*, which was on her first voyage, was an iron barque of 499 tons, owned by W. H. Tucker, and built at Dumbarton in 1875 by McKillay and McMillan. The vessel had been brought back to the U.K. by the remainder of the crew, consisting of James Carrick, A.B., Peter MacGregor, the carpenter, John Dunne, A.B., and two apprentices, Walter Ferguson and James MacDonnell.

The *Cambrian* report (followed by others later, including the *London Illustrated News*, 20th May 1876) filled several columns, and might provide the script for an exciting film scenario. The *Caswell* sailed from Glasgow in July in 1875 with a cargo of iron pipe, under the command of Captain George Best, for Buenos Aires. He was a well-known Swansea shipmaster, who had previously commanded such famous local vessels as the *Swansea*, *Uncas* and *Hawkeye*. His wife was the daughter of Swansea's Chief Constable. It appears that during the voyage to Buenos Aires there had been disputes of a disciplinary nature with crew, and on arrival all were paid off, except for the first and second mates, William Wilson and Allan Maclean, the carpenter, Peter MacGregor, and the two apprentices. In their place Captain Best shipped three Greeks, George Peno, Nicholas Morelie and Christo Baumbos, and two Maltese brothers, Jaspar and Guiseppe Pastoria, together with two British able seamen, James Carrick and John Dunne.

Ashore!

Soon afterwards the vessel sailed for Antofagasta where they loaded a cargo of nitrates for Bristol via Queenstown, Ireland. From the outset, the Greeks and Maltese gave trouble, proving unsatisfactory at their work, insolent and eventually voicing murderous threats against Captain Best. One of the Greeks claimed that he was ill, and when George Peno was ordered to perform his duties, he flatly refused. Captain Best threatened to put him in irons, whereupon Peno, jumped down from the rigging, where he was working at the time, and knifed him, ripping his stomach completely open.

The mate Wilson was running aft, no doubt to procure a revolver, when he was intercepted near the galley by Morelie and Baumbos, who rushed out and stabbed him. They were then joined by the Maltese brothers, Jaspar and Guiseppe, who came up from below, armed with revolvers. (In the subsequent enquiry at Queenstown, Captain Best was criticised for his laxity in not ensuring that the crew were not in possession of firearms.)

The second mate, Allan Mclean, and the steward were then killed, and their bodies thrown into the sea, along with those of the captain and first mate, who apparently were still alive, though dreadfully wounded. The remaining crew would probably have all been murdered but for the fact that they were needed to navigate and handle the ship. Instead terms were made with the survivors, who were sworn to secrecy by the mutineers, relying on their superiority in numbers and arms, on condition that the ship was navigated round Cape Horn, where they could be landed on shore. The carpenter was forced to go into the galley and cook meals, and the remaining British crew, especially the apprentices, were terrorised and frequently beaten, at gun point.

On arrival off Cape Horn, a dispute arose between the Greeks and the Maltese. The former wanted to murder the remaining British crew, while the Pastoria brothers wanted to sail the ship to Buenos Aires, where Jaspar's young wife was then living. One afternoon, Jaspar asked the carpenter, Peter MacGregor, if he could obtain an axe with which to defend himself, because without it he would certainly be murdered. He added that he and his brother meant them no harm; they only wanted to land at Buenos Aires.

A few days later they reached the River Plate, when George Peno announced that the *Caswell* was now a Greek vessel, with himself in command, and ordered the ship's name and port of registry to be painted out, and another name substituted. However, James Carrick and John Dunne managed to arm themselves with an axe and adze respectively, and they succeeded in killing Peno and Morelie as they were climbing up the companion-way.

Next they overpowered Christo Baumbos, though he suffered numerous knife wounds before he surrendered, and placed him in irons. The two Maltese were allowed to escape to Buenos Aires in the ship's boat, and the surviving British crew brought the *Caswell* safely back to Queenstown. James Carrick in particular was singled out for praise, especially as it was recognised that his knowledge of navigation was only what he had been able to acquire while serving in a ship's forecastle. The trial of Baumbos for murder on the high seas was reported in the *Cambrian* newspaper as follows:

> "At the Cork Assizes, on Friday, the trial of Baumbos, the Greek sailor, was proceeded with. – Mr. O'Brien addressed the jury on behalf of the prisoner, and commented that the case resolved itself at the worst into one of manslaughter. – Mr. Murphy was heard in reply for the Crown, and argued that the evidence of the five witnesses for the prosecution was overwhelming proof against the prisoner. – Judge Lawson, in charging the jury, reviewed the evidence, and said a clearer case of wilful and deliberate murder than the present had never occurred. The prisoner would appear to have taken a very prominent part in the whole transaction, and even if he were proved not to have struck a blow, to have been present aiding and abetting, he was equally guilty of murder.
>
> The jury retired at half-past twelve, and in about two hours returned into court having agreed upon a verdict. The announcement caused some surprise. The foreman stated that there was not the slightest chance of agreement. They had agreed with the exception of one juryman, who would not give way. The jury were sent back to their room.
>
> At half-past five they were called out, and stated that there was no prospect of their being able to agree to a verdict. The judge thereupon discharged them, observing that he did not wish to punish by further detention the eleven who were prepared to find a verdict.
>
> At the request of the Crown counsel, it was arranged that the prisoner should be put on his trial again on Monday.

> The second trial of Christos Baumbos for the murder of the officers of the *Caswell* was held on Monday, before Mr. Justice Lawson, in the Cork Criminal Court. The same witnesses were examined as on the first trial, and they repeated the evidence they had previously given. In the cross-examination of Carrick by Mr. O'Brien, the witness became indignant, and said he was advised to summon counsel for having stated that Carrick killed Big George.
>
> The jury retired at 20 minutes past 5 and in 25 minutes returned with a plea of Guilty. The prisoner seemed unmoved and in reply to the usual question of why sentence of death should not be passed upon him, said through the interpreter, Mr. Cartwright, 'Before my God and my country, I declare that I am innocent of the crime imputed to me. It is true that I held one of the mates against the rail, but I did it from fear and compulsion of George, who threatened me with a knife. I deny that I committed any other crime in the vessel. On the contrary, I did my best to save the English crew, the cargo and the ship. This is all I have to say, and I declare before my God, my Saviour and my Redeemer, I did not strike any one with my knife at all."

It would be interesting to know what subsequently became of Carrick and Dunne; they disappear into history, and so far I have not been able to trace any record of them. Should any descendent happen to read this book I trust they will get in touch with me.

THE WRECK OF THE *CATHERINE JENKINS*

The Swansea barque *Catherine Jenkins*, 350 tons, was built in 1845 for William Jenkins of Swansea (see page. 62). Following the death of Mr. Jenkins, his widow sold the vessel to Messrs. Ord of Sunderland who operated a fleet of vessels, mostly engaged in the Swansea copper trade.

The *Catherine Jenkins* was employed on the run to Cuba, the notorious 'graveyard' of Swansea seamen, who so often succumbed to yellow fever contracted there. In Autumn 1855 she sailed from Swansea for Cuba under the command of Captain Matthew Hodge, and made a good passage to Santiago. However, after a few days in harbour more than half the crew contracted yellow fever. Because in those days it was the practice for the crew to load and stow the cargo, the vessel's departure was delayed, but eventually Captain Hodge managed to get away, though with only a minimal crew of eleven on board to sail the ship.

He arrived off Lundy Island on Thursday, 7th February 1856, in bad visibility, but still hoped to make Mumbles Roads by the following morning. However, at about 3 a.m., the second mate heard the sound of breakers ahead, and called

Captain Hodge, but almost immediately the vessel struck the rocks at Lucas Bay, a pebbly cove near Oxwich Point. There was a heavy sea running, but the crew managed to launch the ship's boat, and eleven clambered into it.

They only had two oars, and almost immediately the boat capsized; only four survivors managing to reach the shore. These were the two mates, Hughes and Thompson, and the ship's cook and an ordinary seaman. The captain and four Swansea men, Michael and John Morse, Joe Paine and Ben Austin, were drowned. Ben Austin was the son of Captain Austin, a Swansea pilot. The vessel broke up, and most of the cargo of ore was later carried ashore to a grassy bank, from where it was transported to be smelted in Swansea. The ground where it lay became so poisoned by seepage from the ore, that over a hundred years later its position was clearly marked by its effect on the vegetation on that spot.

THE ALLEGED SCUTTLING OF THE *W. E. GLADSTONE*

The barque *W. E. Gladstone*, 534 tons, of composite construction, was built at Sunderland in 1865, and owned by George Blaney Meager, of Swansea. She was abandoned on the 5th September 1881 off Cape Horn, while bound for Swansea with a cargo of nitrates under the command of Captain Harris, and subsequently became the subject of an enquiry at Swansea as a case of alleged scuttling. The proceedings were extensively reported in the *Cambrian* newspaper, 30th December 1881.

Stories of alleged scuttling were frequent in the Victorian sailing ship era, and many were by no means without foundation. They were difficult, even impossible to prove, given the authority conferred on the Master by the Board of Trade, and the weight given to his entries in the ship's Official Log, especially when these were supported by his officers.

Likewise, shipowners tended to be important members of their local community and were generally well insured, often to the point where the loss of a vessel was a financial gain, rather than a loss. Perhaps a parallel case – hopefully it applied only to a minority – is that of the rascally city businessman, who remarked that 'one more bankruptcy, and I can retire'. Suspicions and gossip invariably intensified when a vessel, to all intents perfectly seaworthy was abandoned in relatively calm weather, not far from land, or in a busy sea lane, where they could expect to be soon rescued.

The case of the *W. E. Gladstone* is an interesting example; there was conflicting evidence on both sides, and nothing was ever proved. The accusation of scuttling came from William Watson, an able seamen, who told the Court of Enquiry in Swansea that he had been ordered below to shift cargo away from the fore-part of the hold. He there noticed that the ship was leaking badly. Before that, an Italian vessel had come in sight, which the *W. E. Gladstone* had soon left behind.

However, later, Captain Harris shortened sail and backed the main yards, until the Italian ship came up to them.

Watson stated that on going below he saw the ship's cook, Michael Holly, chopping at the hull planking on the starboard bow, and asked him what he was doing. The cook replied that 'he was trying to get a hole through her', to which Watson said, 'You are a fool if you do anything of the sort'. The cook's answer was 'I am only doing what I am told'. Watson then went on to say that the carpenter, a man called German, then came down with an adze and began to cut a hole in the planking on the other side. Later he produced a large auger to bore a hole through the ship's side.

Not long afterwards the ship began to settle in the water, and roll heavily with seas washing over the deck beams, and the captain gave the order to abandon ship. All hands were transferred safely to the Italian vessel, lying more or less alongside, and were later landed at Queenstown. Watson there immediately went to the Receiver of Wrecks to report his version of events.

At the enquiry, the mate, David Lewis John claimed that the vessel sank because 'some of the rivets holding the hull planking had rotted away, allowing water to pass in and interact with the saltpetre cargo! At this point the court described Watson as 'a rogue' and told him to 'go away'.

Next the court called the carpenter, cook and another seaman called Musgrave, who all corroborated Watson's account, but claimed they were 'only carrying out orders'. The defence counsel claimed that the true cause of the disaster was that 'a heavy sea had struck the vessel, loosening the hull planking, and this was the explanation entered in the Ship's Official Log. Perhaps not surprisingly, the case ended up as not proven. It is not recorded whether the owner Mr. Meager was present; no doubt he would have been pleased with the captain's vindication.

THE SWANSEA SHIP *STANDARD BEARER*, SCENE OF ALLEGED MURDER ON THE HIGH SEAS 1880; AND LATER LOST OFF THE FALKLANDS 1881

On 20th and 27th February 1880 the *Cambrian* newspaper reported the remand in custody of Captain William Davis, at Bow Street, London, on a charge of murdering a black steward named Price, while commanding the barque *Standard Bearer* on a round voyage to Buenos Aires and Chile. The subsequent enquiry was held in London.

The vessel, 511 tons, and owned by J. E. Burgess of Swansea was built at Barnstaple by William Westacott in 1871 and employed in the Chilean copper trade. She was the last vessel the Westacotts built of any size, and was launched by Miss Williams, the daughter of the local Member of Parliament, C. H. Williams.

The chief witnesses for the prosecution were the First Mate James Craig, the boatswain, James Mclean and William Jones, Carpenter. The vessel left Swansea in March 1879, arriving at Talcahuano, Chile, via Montevideo and Buenos Aires, in September, and after loading sailed for the UK on the 21st October of that year.

The steward, Price, had been shipped at Talcahuano, replacing a previous steward. Three of the crew, Lewis, John and Crean had been imprisoned for refusing to join the vessel, but following an enquiry were released, and ordered to join the vessel. According to Craig and Maclean, Captain Davis was drunk on leaving port, but thereafter conditions on board had been more or less normal, until just before Christmas when an argument had arisen between the mate and the captain over some order relating to the trimming of the sails. The mate had retaliated by saying that on the vessel's arrival in the UK he would report the captain as being frequently in a drunken condition.

On Christmas Day the mate went to his cabin to sleep, but woke up to smell gunpowder. His mouth was full of smoke, though he had heard no report. Putting his hand to his mouth, he found that he was bleeding, and there was a small hole in his lip and a tooth driven out, by a bullet. About eight days later he discovered a piece of the tooth embedded in his tongue. He got dressed, and taking his revolver, approached the captain's quarters, where he could hear the captain and steward talking. He cocked his revolver, and was about to open the door, when he heard a gun shot; he had been shot at the back of the head with a small shot, and he fell down. All the crew were on deck, but after about ten minutes he managed to struggle on deck. He gave over charge of the ship to the boatswain and carpenter, ordering them to secure the captain, which they did.

The following day, another vessel, the *Sardinia* was sighted and signalled, and its captain, William Smith, came aboard with medicines to treat him, and he was able to resume command and bring the *Standard Bearer* back to the UK.

Inspector Butcher of Scotland Yard was then called and stated that he had gone to Dublin to interview Captain Davis, whose defence was that the crew were plotting against him to take his life. He declared that he had overheard the first mate and steward planning to murder him, and he afterwards went to his cabin, where he fell insensible through the fumes of sulphur. The witnesses for the prosecution were unanimous in declaring that Captain Davis was of an unsound mind.

There the account of the trial ends, being adjourned for a further session. Tantalisingly, I have not so far been able to trace the final outcome. As I write this in 1999, I hope to do so in the future. In the meantime, I feel the story is interesting enough as it stands to deserve inclusion.

There is an account in *Sea Breezes* (December 1933, page 269) of the last voyage of the *Standard Bearer* by Mr. S. W. B. Myles of the Falkland Islands who was relating information given by his father. The *Standard Bearer* sailed from Swansea on 4th April 1881 bound for Coquimbo with coal under the command of Captain Manders of North Wales. By the time the vessel arrived off the Falkland Islands it was leaking badly. All wooden vessels leaked, and required constant pumping, but the *Standard Bearer*, built in Devon and not of Canadian softwood, was by no means old. The First Mate, an Irishman named Cody, wanted to put into the Falklands, but at first Captain Manders, who was in his first command, did not wish to do so, fearing his ship might be condemned as unseaworthy. The Falklands at that time had a bad name for over charging for ship repairs, of which, because of their proximity to Cape Horn, there were plenty, and were not averse to generating business by stringent surveys.

The mate had occasion to go down in the forecastle and heard a rush of water beneath his feet. Opening a trapdoor in the deck, to his consternation, every time the vessel plunged in the heavy swell, water rushed in, right up to the breast hooks.

The next morning a vessel, the *Frideburg*, bound for San Francisco, with railway iron came in sight, and Captain Manders and four of his crew went aboard her. The captain of the *Frideburg* refused to take all the crew of fourteen on his ship – owing to lack of space and food, but agreed to take half and give one of his lifeboats to the other half, saying that as they were in the main sea lane, they would soon be picked up. The unlucky half refused to accept that situation, and eventually all the crew boarded the *Frideburg*, taking all the food stores from the *Standard Bearer* with them. The *Standard Bearer* was already sinking fast and was abandoned on the 4th June, the survivors being landed at Port Stanley in the Falkland Islands.

FIRE AT SEA – THE LOSS OF THE *CUBANA*

The *Cubana*, 492 tons, owned by H. J. Madge, sailed from Caldera for Swansea with a cargo of copper ore, under the command of Captain Andrew Peare. According to a report in the *Cambrian* newspaper, 10th October 1862, the vessel, with a crew of seventeen and one passenger was off Cape Horn 54.40 S; 59.25 W when the First Mate, David Hopkins, ran up on deck, half suffocated by smoke, and reported that the vessel was on fire down below.

The flames spread so rapidly that attempts to extinguish them were hopeless, and all on board had to take to the boats, some only partially dressed. It was a stormy night, and snowing heavily; the only provisions they could take with them were about 5 lbs of biscuits, 4 gallons of water, and some fresh pork. The smaller of the two boats, a pinnace, was swamped but those aboard managed to scramble into the other boat.

After seven days of intense privation and cold, they reached a small uninhabited island to the south of the Falklands. By that time only two of them were able to walk. The island was covered with snow, but they managed to get the boat up on to the rocky shore, and spread a sail over it to provide a shelter. There they remained for eleven days, living on shellfish and penguin meat. By this time nine of the survivors were in a terrible state with their hands and legs severely frost bitten. It was decided to leave them, with two men to look after them, while the others put to sea in the boat, hoping to reach the main Falkland Islands. After two days they were picked up by the schooner *Perseverance* of Port Stanley, belonging to the Falkland Island Company, which was on a sealing voyage. Another schooner, the *Victoria*, picked up the others from the island, and all were taken to Port Stanley for hospital treatment. The Second Mate, George Elliot, and one of the seamen died shortly after their arrival, and four of the seamen had to have their legs amputated.

Even at this length of time it is painful to relate such harrowing details. The men who sailed in the Swansea copper barques must surely be unequalled for their hardihood, courage and seamanship.

THE *HECLA* – YELLOW FEVER, AND LOSS AT PUNTA BLANCA

The barque *Hecla*, 447 tons, was built at Hartlepool in 1854 and although owned in Sunderland by Ord & Co., was regularly engaged in the Swansea copper ore trade to Cuba, and later, Chile.

The *Hecla* was notorious as the ship that brought yellow fever to Swansea in September 1865 when she anchored at Mumbles Roadstead with her flag flying at half-mast. The pilot brought news that five of the crew had died of yellow fever contracted in Cuba, and fresh hands were sent out to bring the vessel into port – surprisingly in the circumstances. One seaman was carried ashore in a litter and died of yellow fever a few hours later.

Subsequently there were many recriminations over the rash action of bringing such a contagious vessel into the port. A special meeting of magistrates was convened 'to prevent any evil consequence arising from bringing such a vessel into the port'. At a meeting of the local Board of Health the mayor explained that as soon as he heard of the arrival in port of the *Hecla* 'he at once saw the owner of the vessel, and others and had the ship, the clothes of the crew, and the house to which the dead seaman had been removed, thoroughly fumigated'. He had also 'seen the officers of the customs and the coastguard, and could now guarantee that no other vessel under sail or circumstances would be allowed to enter the port'. (The *Cambrian* Newspaper, 15th September 1865.)

The career of the *Hecla* ended during a passage from Valparaiso to Tocopilla, when on 18th November 1870, the captain, James Rudman thought he was in the vicinity of Tocopilla, when he was in fact off Punta Blanca. He only discovered this as he was, as he thought, about to enter port, and a local boat, hailed him with information. While endeavouring to save his ship by heading out to sea, the vessel struck a submerged rock, and was so severely damaged that it began to break up. The crew of sixteen took to the boats, and all landed safely at Punta Blanca, from where they were sent to Valparaiso. Subsequently the British Consul arranged for their passage back to the UK.

At the subsequent enquiry in Swansea, Thomas Robson, the boatswain and acting Second Mate declared that Captain Rudman had not inspired his crew with confidence, and left the first mate to do most of the navigating. Captain Rundman's own explanation was that the shore buildings at Punta Blanca closely resembled those he had been looking for at Tocopilla, i.e. the various buildings and chimneys associated with the smelting works, together with a British flag hoist. He therefore concluded that he was off the correct port, and his ship consequently in no danger. He was a fortunate man, for he was exonerated of any positive neglect, and did not lose his certificate of competency.

THE *CALDERA* – COLLISION OFF CAPE HORN

The *Caldera* was an iron barque, 469 tons, built at Sunderland in 1861, and owned by H. J. Madge of Swansea. In March 1873, while bound from Chile to Swansea with copper ore under the command of Captain William Griffiths, she was in collision with the *Canning* of London about 800 miles from the Falkland Islands.

According to Captain Griffith's account, later accepted by the Court of Enquiry: the *Caldera* was going free when the *Canning* was sighted from 1½ to 2 points on her starboard bow, there being no light aboard the *Canning* she was deemed to be going west, and the helm of the *Caldera* was starboarded, to give her more room. When the green light of the *Caldera* was seen, the mate of the *Canning* immediately ported, and the result was collision.

The Court therefore ruled that the want of lights on board the *Canning* was the primary cause of the disaster, and that the mate of the *Canning* was not justified in porting to a green light. They also ruled that it was justified in coming to the conclusion that 'ours was the following ship, as there was nothing to guide me, there being no lights on the *Canning* and the failure to obey the law led to the accident they must adjudge the *Canning* to be wholly at fault'.

The damage to the *Caldera* was severe; the whole upper works were 'broken' and the first and second mates lost in the disaster. It was clear that the *Caldera* was in parlous condition, and Captain Griffiths decided to try and sail his ship to the Falkland Islands, pumping continuously. He succeeded in getting near to Port William where he dropped anchor, and when it was clear that the crew could safely make shore, Captain Griffiths set fire to his sinking vessel, in order to prevent her staying afloat as a derelict and therefore becoming a hazard to shipping on a busy sea lane.

He was at first criticised for this, and even for attempting to make for the Falklands with a disabled ship without a first and second mate. His defence to the Court was: 'Now as to the last act of firing, in that I might have erred; but if so, it was on the side of humanity . . . what sad disasters occur off the Horn, through the elements and the seas, without having derelict ships add to their number.' In fact at the first hearing his certificate was suspended, but at a second hearing the Court, after further reviewing the case, decided that his actions in the circumstances were correct, and it was returned to him, with no stain upon his character.

LOSS OF THE *GUAYACAN*

The *Guyacan*, a 451-ton barque built at Sunderland in 1857, was owned by Henry J. Madge of Swansea and lost at Trinidad on the 20th February 1865. An account in the *Cambrian*, 10th November 1865, fills five columns from which I have extracted the following:

> "She left Carazal on the 21st December last, in charge of Nathaniel Frances Carder – her master – who had a certificate of competency, but which, together with the ship's log-book, was lost. The ship had a crew of 14 hands all told, and she was then laden with about 700 tons of copper regulus. He need say little of her voyage beyond that she rounded the Horn successfully, and about the beginning of January she got into the trade winds, which generally speaking were northerly, varying from North to East and E.S.E. The vessel was sailing before the wind the

'Aberlemno' aground at Ilfracombe, 1899 (possibly the 'Serena').
(National Maritime Museum, Greenwich)

The barque 'Kelburn' in port after going ashore (subsequently scrapped).
(National Maritime Museum, Greenwich)

Saturday before she struck and observations were taken. The vessel struck on Monday the 20th February last, when she was in lat. about 12.40S. and long (by dead reckoning) about 35.40 W. The day before she was struck, and about the time she did strike, the wind was about E. by N. half N. to E., and the vessel was going about six and a half knots. It was about 20 minutes past 12 a.m. on the morning of the 20th February that the vessel grounded, and it was whilst she was in charge of the chief officer who had been on watch since midnight (only, in fact, 20 minutes before the vessel struck). There did not appear to have been any previous signs of breakers ahead or anything whatever to warn the crew of their danger – the striking was quite sudden and altogether unexpected. As soon as the ship struck the mate ordered the helm to be put down, and then called the master from the cabin. The yards were then ordered to be backed and the staysail hauled down, and also the fore and aft sails, and the vessel came up to N.E. and fell off before the wind, banging by the stern, which was fixed in the sand-bank. The vessel appears to have drifted some little distance along the sand, in consequence of which a stream anchor was taken astern, with a cable attached to the vessel, which held her steady. Within an hour after she struck the carpenter found about eighteen inches of water in the hold, and early next morning about 10 feet of water. That morning a boat was despatched to the nearest port for assistance, and the British Vice-Consul came on board, and at once advised them to do what they could to save the cargo. The vessel soon after began to break up, but between 300 and 400 tons of the cargo were saved. The ship became a total wreck, and now lies buried in the sands. He was happy to say no lives were lost."

The following evidence was then adduced:

"J. Boden was first witness. He said: I was chief mate of the *Guayacan*, holding a certificate of competency for the first mate, dated 11th April, 1864. We left Carazal on the 20th or 21st December, 1864. The log-book was lost. The time the ship filled I was on shore putting the Consul on shore. All the crew had left when I returned except the captain. I left the log-book on my desk in my berth when I left the ship. When I returned the vessel was full of water. I never saw the log-book after, it must be under water.

Edward Birch was the second mate, but he was off duty. He was in my watch and was on the deck when the ship struck. The boatswain kept the first watch. The second mate was a certificated mate. He was off duty because he did not suit the captain – he had been put off duty by the captain on the voyage out, and was not restored. The Consul had the official log-book.

Captain Harris: There is nothing in the official log about Birch. We are at a loss why he was off duty.

Examination continued: It was about the latter end of August, 1864, the mate was put off duty. We left Swansea on the 5th July. The reason the mate was put off duty was because he did not handle the ship properly – he was put before the mast – disrated. He continued doing his duty as an able seaman up to the loss of the ship. He was placed in my watch. The master was on deck about ten o'clock the night before the ship struck, trying to take observations by the stars. I was lying down on the hen-coop on deck. I do not know whether he succeeded in getting an observation. The ship struck about twenty minutes past midnight. I was then on the poop, looking over the lea side towards the land. We did not expect to see land – nobody saw the land when the ship struck. We first saw the land about an hour after the vessel struck. It was a low sandy beach. The look-outs did not report any land. They did not report anything before the ship struck. I did not see any roll of the sea or broken water before the ship struck. It was very dark at the time. I did not notice any discolouration of the water. No lead was hove before she struck. the vessel struck aft first – very easily – it was smooth water. The helm was then put down by my orders, and the yards put back. I called the master, and before I had the words, 'Captain, the ship's ashore,' he was on deck, as he felt the shock in the cabin. The ship came up about a point, when the helm was put down, but she lost her way and went back again before the wind with her head towards the shore, and remained fast.

Mr. H. J. Madge sworn: I am the managing owner of the *Guyacan*, and of several other ships. I have been personally acquainted with Captain Carder for the last seventeen years – he has been in command of ships in our firm for the past fifteen years; as the vessels have changed or renewed he has had the preference given him. His character has been very good in every respect, as to seamanship and as to keeping the accounts, and also as to sobriety. He has been habitually a sober man. I have never seen him anything the worse for liquor. I have always known him as a remarkably temperate man.

By Captain Harris: Captain Carder has been engaged in the west coast or copper ore trade during the whole of the time he has been in our employ. We have never had any occasion to remonstrate with him for too closely hugging the land. As managing owner of the ships of our firm, I am perfectly satisfied as to the way he has managed our vessels. Captain Carder was part owner of this ship. He has frequently been in the habit of keeping closer to the Brazilian coast on the outward voyage than is usually the case, but not on the homeward passage.

The Court then retired, and after an absence of about a quarter of an hour, Major Francis delivered the following judgement: The Court having given due consideration to the evidence in this case, is of opinion that the barque *Guayacan* was carelessly navigated, and that she was lost by the default of the master, Nathaniel Francis Carder, inasmuch as he hugged the coast too closely, although his course and some indications of land ought to have warned him of danger. The omission to throw the lead in such circumstances was unwarrantable and the Court feels bound to draw special attention to this grave omission. Taking all the circumstances into account, the Court feels constrained to suspend the certificate of competency of the said master, Nathaniel Frances Carder, and it is hereby suspended for the space of six calendar months."

ALLEGED MURDER ON THE *SUNBEAM*

The *Sunbeam* was a brigantine, 284 tons, built at Prince Edward Island in 1880 by John Yeo, and owned by John James Dale, a timber dealer of Swansea, and J. C. Richardson. She was eventually lost near Puerto Coballo, 28th March 1884, but in 1881 was the scene of an alleged murder that was reported at length in the *Cambrian*, 29th May and 17th June 1817. The first report was as follows:

"From particulars published in the South Carolina papers, the *Sunbeam*, a vessel owned in Swansea, has got into unenviable notoriety owing to the extraordinary conduct of her captain, who is at present awaiting his trial on a charge of murder. The account of the affray is as follows: It appears that some time ago arrangements had been made with a man named Sweeney to supply the *Sunbeam* with four sailors. On Saturday last he arrived in Beaufort, and the men with him caroused about the bar-rooms in town, and before leaving for Bull River were considerably the worse for liquor. Arriving on board on Sunday morning the sailors got into a dispute with a coloured watchman, and commenced a fight in the forecastle, which extended to the deck of the vessel. The Captain sent for Sweeney and told him to take the men off, as he did not want such cut-throats on his vessel. Sweeney told him that he would quiet them, and that when they got sober they would be all right. Sweeney went forward and quelled the disturbance. Soon after this the captain paid him the advance money for the men, and about eleven o'clock Sweeney took to his boat to return to Beaufort. The sailor Drummond went over the side of the ship and joined Sweeney in the boat and was followed by Murphy. The captain, fearing they intended desertion, pulled out his revolver and snapped it three or four times, but it missed fire. He detained the boat by seizing hold of the jib stay, and told his boatswain to bring him his gun. This was done, and the captain emptied the contents of one barrel into the left side of sailor Murphy. It was loaded with 6 duck shot, which penetrated the sailor's heart, causing instant death. At this time Murphy was attempting to cut the boat's stay. The Captain then turned upon Sweeney, saying, You are the next man. If you were never shot before you will be to-day. He then fired, the shot taking effect in the left arm of Sweeney, causing a painful although not serious wound. The sailor Drummond was also stuck on the head with some of the shot and slightly wounded. The boat drifted from the ship, and the captain fired four shots from his revolver after it. The men in the boat hoisted sail, and arrived in Beaufort with the dead and wounded about four o'clock Sunday afternoon. Coroner Burr was telegraphed for Sunday evening, and early Monday morning empanelled a jury, who returned a verdict of feloniously killing against Captain Thomas. The same evening the sheriff arrived with the captain and the boatswain in custody, and lodged them in gaol. The witnesses are also detained in custody. No fatal result is anticipated in the cases of the wounded men. Colonel William Elliot has been retained to defend Captain Thomas, who, we understand, under advice has made no statement for publication."

The account of the trail filled several columns on the 17th June and the following extracts could almost form a film scenario:

"In brief, the witnesses for the defence testified, in substance, that Sweeney was a desperate character, that he had been engaged for some time in the business of enticing sailors away from ships after they had been articled, and after he had drawn their advance money to liquidate the real or imaginary claims he held against them; that Murphy was also an unruly ruffian, and had long been with Sweeney and others in the nefarious business they were engaged in, and was merely a runner for a sailor boarding house; that McKain was what is known among seafaring men as a 'nautical tramp', who travelled from port to port 'beating' or assisting to 'beat' shipmasters, and that Drummond was formerly a stevedore, but had left that honourable occupation to follow the apparently more lucrative one of 'jumping advances', that the three parties had signed themselves, or had had signed for them, assumed names upon the shipping papers; that Murphy always carried upon his person a sheath knife of ample proportions: that when the party arrived at Beaufort they offered to pay Simon Wethers not less than eight dollars each to be taken to the ship and brought back; that Wethers refused because he was afraid trouble would come of such an undertaking; that they then hired James Smith to perform that duty for ten dollars each; that they were very drunk, or apparently so, when they left Beaufort for the barkentine; that they boarded the *Sunbeam* with no baggage except for a bed of straw each, with the exception of the fourth man, a coloured man named Hull, who carried his 'kit', and went on the voyage with the *Sunbeam*; that no sooner were they aboard than they picked a quarrel with a coloured man, who was an old hand on the ship, and beat him severely in the forecastle; that the coloured man escaped from them and fled to the cabin; that the captain and the mate came up to quell the outbreak, and were violently threatened by Murphy and Drummond, the former of whom struck down and booted the Negro brutally abot the face and head; that it was impossible to restrain the rioters in their violence until Sweeney bade them wait; that shortly everything quieted, and the crew retired to their quarters, leaving the deck bloody in many places from the fracas which had taken place, and the injured man suffering from his bruises; that Sweeney returned to the cabin and concluded the money arrangements he was making with the captain, after which he went into the forecastle and had a long confidential conversation with the three white sailors who had come aboard with him; that he came forth, bade them goodbye, ordered Smith to get aboard the small boat, and walked aft to the same; that after he was in the boat Drummond and the others came aft and started to get into the boat 'to shake hands'; that Captain Thomas ordered them to their quarters; that they persisted in their design and sprang aboard; that Smith and Sweeney had drawn the boat along the side of the vessel much nearer to the bow than it lay when the 'painter' was first cast off; that an ordinary man standing erect in the small boat would have his head and shoulders above the rail of the *Sunbeam*; that the *Sunbeam* was loaded and ready for sea, and lay very low in the water; that the captain grasped the iron jibstay and ordered the men to return on board, threatening to shoot them if they disobeyed; that they cursed him, and he snapped an empty pistol at them; that they still refused obedience, and he called for his gun; that the boatswain brought a shot gun at full cock, which had been loaded that day with small shot to shoot birds; that the captain had the jibstay with his right hand; that the gun was taken by him with his left hand and rested upon the rail; that all this time the captain was calling upon the men to return; that Murphy sprang to the bow of the small boat and slashed at the captain with his knife; that while slashing and cutting at the captain, and resting his knee upon the side of the barque, the captain shot him with one barrel of the fowling piece, and he desisted from his efforts; that the captain then observing Sweeney draw his pistol and level it at him, fired the other barrel to disarm him, the shot taking effect in Sweeney's right arm; that the men in the small boat huddled on the bottom of the same, and it floated some distance from the ship with the side toward the ship careened up so as to hide the occupants from the range of the captain, who fired his revolver several times in the air to intimidate them, he not knowing at that time that his first shot had proved a fatal wound; that Sweeney, when leaving Charleston with the men, told a party that he was going to Bull River to beat a 'lime-juicer' out of an advance; that the captain, as soon as the boat passed away from the barque prepared a statement in writing of the whole affair, and requested Pilot John Murray, who was aboard to transmit it immediately to Sheriff Wilson, of Beaufort, which was done.

W. J. Verdier Esq., one of the attornies for the defence, then made a most exhaustive and convincing argument upon the evidence, going over the entire ground with careful and deliberate steps, and closing with a most powerful appeal to the jury in behalf of his clients.

The jury remained out about fifteen anxious minutes. On their return a verdict of 'not guilty' was rendered. It was almost impossible to restrain the populace, who seemed to have sympathised deeply with the defendants throughout the trial, and upon the discharge of the prisoners their hands were warmly wrung by many."

THE RESCUE BY THE SWANSEA BARQUE *CUBA* OF THE CREW OF THE *BONA DEA*

This was reported in the *Cambrian*, February 17th, 1854, as follows:

Dreadful Shipwreck, & Horrible Sufferings of the Crew

By the arrival of the barque *Cuba* at Swansea this week, we are put into possession of the details of one of the most heartrending shipwrecks it has ever been our lot to record. The fullest particulars will be found in the subjoined statements:

Captain Organ's statement

On Saturday, February 4th, at about 10 a.m. (being then in lat. 45° 15' N., and long. 24° 14' W.), saw a vessel about two or three miles to windward, apparently in distress. I immediately shortened sail, and hauled up for her, and made her out to be the *Bona Dea*, of Liverpool, with only fore-mast and fore topmast standing, and water-logged – a perfect wreck, and the sea making a breach right over her. I then discovered that there were people still on board, and immediately got a boat out and sent to her, in charge of the chief mate, to take them off the wreck; which, with some difficulty, they succeeded in doing. When we got them all on board the *Cuba* (eleven in number), we found them in a truly wretched and exhausted condition, and presenting a most melancholy and distressing sight, having been eleven days and a half in that fearful situation, without either food or water. One poor fellow (Samuel Blane, cook) died almost immediately after we got him on board; four of the others were quite insane. We immediately proceeded to give them a little boiled rice, barley water, &c., feeding them with a few spoonfuls; we then got them all below, and washed them with warm water, and put clean dry clothes on them, and after we had made them as comfortable as we possibly could, some of them appeared to revive a little; but those who were insane appeared to get worse, especially one of them (James Gilmour, carpenter), who, after raving part of the night in a most furious manner, at last sank, exhausted. All our efforts to restore him were in vain, and, after lingering all the next day (Sunday, 5th), at 8 p.m., he expired. Three of the others were very delirious, but we could get them to swallow small quantities of food, barley water, &c. On Monday, the 6th, two of the three who were delirious were fast sinking, and past all hopes of recovery; all the others appeared doing as well as we could expect, and very ravenous for food, but we were cautious not to let them have more than small quantities at a time, but at frequent intervals, such as weak broth, rice, gruel, &c.; at 3 p.m., one of the two who was so far gone, expired (George Nicholas, an Austrian), the other one scarcely breathing. On Tuesday, the 7th, Robert Anna, the only one man who appeared past recovery, was still breathing, but could not get any thing down his throat; he was too far gone. All the others appeared to be still improving – one, who was quite delirious when he came on board, had now some lucid intervals, and would talk quite reasonable, and be conscious of every thing. They complain principally now of their feet and legs, which are much swollen, and very sore. On Wednesday, the 8th, at abour 2 p.m., Robt. Anna expired, after lingering since yesterday, scarcely breathing. I committed his body to the deep that evening, and read the burial service, which I did also when we buried the others, respectively. On Thursday, the 9th, they all appear to be recovering fast, and I have now good hopes that they will recover, but I feel persuaded, that very few of them could have survived another day on the wreck, judging from the wretched state they were in when we fell in with them. And I shall never forget the distressing sight those poor emaciated, famishing creatures presented; and, while I thank Almighty God for being the means, in his hands, of rescuing so many fellow-creatures from a horrible and frightful death, I trust I may be spared from ever witnessing such another awful scene.

Up to the time of our arrival at this port, the seven survivors were daily improving, and I consider them as now all quite out of danger; although still weak, and their feet and legs very sore, and still much swollen.

The following are the names of those who came on board the *Cuba*:

<table>
<tr><td>Samuel Blane, cook
James Gilmour, carpenter
George Nicholas, seaman
Robert Anna, seaman</td><td>Died on board the Cuba, and buried at sea.</td></tr>
<tr><td>Duncan McLeod, chief mate
Thomas Phillips, second mate
John Doherty, seaman
Benjamin Mooney, seaman
William Harrison, seaman
John Naylor, seaman
Charles Philpotts, seaman</td><td>Survivors arrived at Swansea in the barque Cuba.</td></tr>
</table>

Signed, FREDERICK JOHN ORGAN,
Master of the Barque *Cuba*, of Sunderland.
Swansea, February 15th, 1854.

EXTRACT FROM A STATEMENT MADE BY MR. MACLEOD, CHIEF OFFICER OF THE *BONA DEA*

. . . the watch being just called I was preparing to go on deck, when a heavy sea struck the ship on the starboard quarter, causing one of the heaviest shocks I ever experienced, and fairly lifted me off my feet. The pumps had just been sucked by the watch on deck. My watch relieved them, and continued pumping without being able to get them to suck. We were now running under close reefed fore topsail and fore topmast staysail, having split the foresail the same day. At 10 p.m. the fore topsail blew away. We then hove the ship to under the main trysail, with the head hauled down. The ship was now labouring heavily and making fearful weather. At midnight a heavy sea broke on board, taking away the larboard bulwarks and stanchions, fore and aft; breaking adrift the deck load; hove several bales of cotton overboard which were floating about adrift. All hands lashed to the pumps (myself included); we continued pumping till about three a.m., when, not being able to get her to suck, the carpenter was sent to ascertain what water was in the ship, and found she had filled to within three feet of the main deck. The captain then called all hands away from the pumps, and made known to them the state of the ship, and there being no time to lose, he ordered me to take some hands and endeavour to get a cask of water on the poop, while he himself attended to the getting some provisions out of the cabin. This was the last time I saw him, for shortly after (before we could succeed in getting the water and provisions on the poop), the ship gave a heavy lurch, and capsized right over on her broadside. She lay on her beam ends for some time (about an hour), when those of us who were saved cut away the main and mizen shrouds; being on the broadside, and sea breaking right over us. When the masts went she righted again. We then began to look round and enquire for each other, and found that the captain, sailmaker, steward, an ordinary seaman, and a boy, were missing, and we were now only twelve in number out of seventeen; we also found that everything had been taken away, having washed everything out of the cabin, forecastle, and off the deck – we had not a single thing left – not one morsel of bread or a drop of water. The mizen mast had gone by the saddle. We managed to get an old sail over the stump of the mast, and lashed ourselves there, expecting every moment the poop would go to pieces, as two of the main beams of it were gone. We remained in that situation all that day.

Tuesday, January 24. Still blowing as hard as ever, and the sea washing over all; we wished to get into the fore-top, but it was impossible to get along the main deck. I did all in my power to endeavour to cheer up the rest with the hopes of soon being relieved, which by God's help, I in a great measure succeeded in doing even to almost the very last. We kept under the temporary tent we had made over the stump of the mizen mast, but kept a sharp look out night and day.

We succeeded in cutting away the deck, and could just reach the casks, but were not strong enough to get them out and after exhausting what little strength we had, we were obliged to give it up in despair. The carpenter, raving mad, and holding on to the skylight, and would not let it go; the cook, and Geo. Nicholas (seaman), also quite deranged – these three had been out all night exposed to the weather, having refused to come under the tent; Charles Philpott, and Robert Anna (seamen), very delirious inside the tent, and completely helpless; the others were sensible, but quite exhausted; myself and two others just able to walk. Such was our state this morning, when, at about 9 a.m., I saw a vessel through the haze about five or six miles off, running eastwards, and it would be impossible to attempt to describe our feelings at this moment (those few of us who were still conscious of any thing), we even commenced hailing. In about an hour, saw her shorten sail, and haul up for us; and when near enough hove to and sent a boat to us, in charge of the chief officer, who managed it with much prudence and skill, as it was both difficult and dangerous coming alongside the wreck, and required much caution as the sea was constantly washing over the wreck; they were obliged to make two trips with the boat, to get more of their own crew, to get those who were helpless and insane into the boat, and being obliged to lower them down with ropes, which made it very dangerous, as it was difficult to keep a boat alongside the wreck; but they succeeded in getting them all off the wreck, and at about 1 p.m. we were all on board the barque, which proved to be the *Cuba*, from Coquimbo, and bound to Swansea, F. J. Organ, master; she lost about three hours of a fine fair wind from the time she first hauled up for us, until she kept away again.

The cook, as soon as he was taken on board, was taken down in the cabin and every means tried to restore him, but all in vain; he expired almost immediately. He was washed, sewed up, and buried next morning in a respectful manner, according to the rites of the Church of England. All the rest of us were washed clean and nursed, as though by our own relatives – put clean and dry clothes on us all, and beds to lay in. Captain Organ and all his crew attended to our wants, and in truth they had much patience to endure our unreasonable requests, which, if compiled with, would have done us serious injury in the state we were; but Capt. Organ understood our case better than ourselves, and would only let us have small quantities of food at a time at first, and increasing the quantities as we were able to stand it; and such as rice, sago, barley water, &c. Nothing was spared that was in the ship – even the preserved provisions, which were so much needed for themselves coming such a long voyage, were used exclusively for us, to bring us round. Suffice it to say, that such kind care and treatment stands unequalled from strangers, as we have received in this truly Christian ship. The carpenter died on Sunday, Feb. 5th, quite insane, and was buried on Monday. Captain Organ read the burial service over him. George Nicholas, an Austrian, died on Monday, Feb. 6th, and was buried next day in the same manner. Robert Anna died on Wednesday afternoon, and was buried that evening, Feb. 8th. All these had been quite insane. On each occasion all the crew of the *Cuba* were in attendance in a most respectful manner, and showed more Christian feelings and conduct than I have seen on any such occasions. There was still one more delirious, but getting better. All the rest of us getting on well, and better than might be expected from the state we were in when we came on board. We were long and anxiously looking for relief, and we may thank God that we fell in with such a ship as this; had we fallen in with indifferent people, I feel sure there would not be many of us left. We were watched and nursed night and day as though we were their own brothers. They had, indeed, much trouble with us, to bring us to our present condition.

The men were divided, those who were insane were put in a roomy, clean, and comfortable aft. deck, the others were in the forecastle with plenty of room, a fire and everything suitable for them, and myself in a comfortable cabin, with a fire, a good bed, and every convenience and attention that man could desire. And all I can do or say, for this unheard of kindness from entire strangers, is to pray to that merciful God, who has spared and delivered us, that none of them may ever experience what we suffered on board the *Bona Dea*.

(Signed) DUNCAN MCLEOD,
Mate of the late barque *Bona Dea*.

[It is but justice to Capt. Organ to state that his conduct on this trying occasion was not only humane, but that of a true British seaman, animated by the best feelings of our nature. Nor was his benevolence confined to his own ship. Since his arrival in port, everything has been done by him to promote the comfort of the men, and we are happy to hear that with the aid of Capt. Edwards, our harbour master, they have all had a passage to Liverpool in the *Glendower* steamer.]

I cannot end this chapter without mentioning the epic voyage of the *Merioneth*, a very ordinary Welsh barque which in 1888 took a cargo of coal from Cardiff to San Francisco in 96 days. It was a record that was never beaten, even by the clipper ships, and of an estimated ten thousand voyages from the UK to San Francisco it is believed that only twenty-five completed the voyage in less than one hundred days.

The *Merioneth*, commanded by Robert Thomas, was a North Wales ship, but I am sure our friends in North Wales will not begrudge a mention of her tremendous achievement in a book about South Wales ships and seamen.

I give below an extract from an article which appeared in the San Francisco *Daily Journal of Commerce*, March 2nd 1888, to celebrate the achievement of Captain Thomas and his crew:

> ". . . but now at least, Uncle Sam must confess that his best effort has been outstripped by Britain, or at least by old Gwalia, for the *Merioneth*, as her name would indicate, is a Welsh ship, with a Welsh commander, a Welsh crew, and owned by a Welshman . . .
>
> In recognition of the splendid seamanship displayed by her commander, Messrs. J. D. Spreckles & Bros. have presented him with a beautiful silk flag, which as long as it lasts, will commemorate the memory of this remarkable voyage."

Captain Robert Thomas of the 'Merioneth'.
(Gwynedd Archives Service)

The 'Merioneth' (in centre) at San Francisco in 1888 after her record-breaking voyage.
(Courtesy San Francisco Maritime Museum)

The crew of the 'Merioneth' at San Francisco.

Captain Thomas died in San Francisco in the Spring of 1903 after an epic voyage around Cape Horn in command of the barque *Afon Alaw*. His daughter was aboard, and from her letter home describing the funeral comes this description:

> "Behind the hearse walked the pastor and pall-bearers, next came the sailor boys with flowers, next came the carriage with us, and behind us all the women and shore friends, a great long line stretching down the street, and after these came all the captains . . .
>
> Arriving at the cemetery the coffin was born by the cemetery workers and just as they halted one of the bearers turning round and seeing all the long straggling line of sailors behind, asked most excitedly, 'My God, who am I carrying?' "

11.

The Other Side: Cuba, Chile, Peru and San Francisco

CUBA

THE FIRST FOREIGN COPPER ORE to reach Swansea in major quantities came from Cuba, where it was mined at Cobre and shipped from Santiago de Cuba at the south west of the island. The Cobre mine had been originally opened by the Spanish and then abandoned in the 1720s, but an Englishman who visited Cuba in the 1830s to recover a mortgage debt assayed some waste from the mine which turned out to be extremely rich. The mine, now English-owned was re-opened with Cornish miners, a large number of whom were soon struck down by the yellow fever endemic in the island. They were replaced by local labour, less liable to die from the fever, as well as labour from the Canary Islands. David Turnbull, visiting Cuba in 1838, described the mine-workers as little better than slaves.[1]

At Cobre the ore was stamped into pieces by steam-hammer and loaded into baskets for a twelve mile journey by mule or horse into Santiago. At the other end in Swansea, the Cobre Company had its own wharf, the Cobre in the North Dock.

Yellow 'Jack' took a terrible toll of Swansea seamen and Santiago was referred to as 'Swansea Cemetery' because of the large numbers of Swansea men who were buried there. The pages of the Swansea newspaper, *Cambrian*, are filled with reports of ships arriving at the port having suffered crew losses from 'Yellow Jack'. Susan Campbell-Jones in her *Welsh Sail* mentions a register kept in Santiago in 1862 listing the deaths that occurred on Swansea vessels during that year:

San Jose	2
Florence	1
Cornwall	10 (5 at Cuba; 5 at sea)
Ellen	8 (7 at Cuba; one on the way home)
Mangosteen	1
Hampshire	1

1. Ronald Rees (New Brunswick, Canada), *'King Copper': South Wales and the Copper Trade, 1584-1895.*

Countess of Bective	5 from yellow fever; 1 from brain fever
Dorsetshire	4
Cobrero	4

Black cooks were often carried on vessels trading to Cuba since the exposure of a European cook to the heat of the galley seemed to increase the chance of contracting yellow fever. Under Spanish law, free blacks and mulattos were not allowed to land in Cuba and it was necessary for the Master to make special arrangements to allow them to do so.

In January 1845, twelve barques arrived at Swansea from Cuba within a fortnight, every vessel having lost at least one of her crew from fever. On the *Lady Pirie*, six had died. There were several cases where the port quarantine regulations were not observed, and the disease was carried ashore, where it spread through the town. The worst example of this was when the barque *Hecla* arrived at Swansea on September 8th 1865. At first the captain had tried to conceal the presence of the disease, and within three weeks, fifteen Swansea people had died (see page 167).

CHILE

The first ships to visit the west coast of South America were the treasure galleons of the Spanish, which were the prey of British seamen like Drake and Anson. A hundred years afterwards they were followed by the first of the Swansea copper barques, and later again by the Guano ships and finally the nitrate clippers.

A glance at the map shows Chile as an extraordinarily long thin country which stretches more than half way up the west coast of South America between the Andes and the sea. Its people are of mixed Indian and Spanish descent, of whom only a small proportion are pure-blooded Indian, and all are Spanish-speaking.

The capital is Santiago, roughly half way up the coast and about fifty miles inland from Valparaiso, the main port. The country's copper industry is still today the fourth largest in the world, and nitrates and molybdenum are nearly as important. Chile was also famous for its wine industry, which in recent years has rapidly developed into a large export trade.

North of Valparaiso lie the famous ports which occur again and again in the ship's logs and records of Swansea vessels such as Coquimbo, La Serena, Antofagasta, Huasco, Caldera, Tocopilla, Iquique, Pisagua and Tacna. To the south there was Talcahuano, Conception and Valdivia.

Their names roll off the tongue, reminding those of us who are familiar with such things of the famous Swansea vessels which were named after them. We reflect that those far-off places were familiar topics in the public houses which bordered on the Strand, and in the homes of Swansea seamen.

Chile is well-known for its proneness to earthquakes; in 1866 and 1877, Iquique was nearly destroyed by them, and on the last occasion the destruction was com-

The anchorage, Pisagua.
(Nautical Photo Agency/National Maritime Museum, Greenwich)

pounded by a huge tidal wave. The Valparaiso area has suffered similarly and there is in Swansea an interesting link with the church of La Compania at Santiago originally built in the 16th century and destroyed and rebuilt following two earthquakes, the last being in 1857.

On the 8th December 1863, when the church was full to capacity and filled with candles and paper decorations, fire broke out, and within an hour engulfed the building. In the ensuing panic the exits became blocked, and nearly 2,000 people lost their lives. When the tower and belfry crashed to the ground, the great copper bells which they contained fell with them.

The story goes that Graham Vivian, the younger brother of Henry Hussey Vivian, was in Valparaiso at the time on the firm's business. He must have gone up to Santiago, for he saw the bells, and seeing the fine ornamentation and inscriptions on them, arranged to acquire them. It is said that the bargain was to exchange them, weight for weight, with the existing bells in All Saints Church, Oystermouth. Whatever financial arrangements were made, the bells were brought back to Mumbles, where they remain in the church today.[2]

Chilean ore was not only plentiful but of very high quality, and for thirty years it formed the main supply of the Swansea smelters. The organising centre was at Valparaiso, where the barques usually called for orders, and then went to the other

2. Susan Rees, *In Search of the Bells of Santiago*.

Off Valparaiso, 1868. Painting by James Harris Jnr.
(Private Collection)

ports to fill up. The nearest mines to Valparaiso were at Coquimbo and La Serena, a few days passage to the north. Coquimbo possessed a safe bay and a good anchorage, and the river Coquimbo, on which both towns stood, flowed through a fertile valley.

The copper ore was extracted from the mines by local labour in conditions that were horrific. Sacks weighing up to two hundred pounds were carried up ladders to the surface on the backs of straining labourers, and it was only when the mine was more than 600 feet deep that they were allowed to rest on the way up.

Vivian Webber visited Charles Lambert's mine at Coquimbo and in his journal he described his experience:

> "Having divested ourselves of our coats and put on a sort of blouse, we entered the mine by what is termed a horizontal shaft; after walking for a quarter of a mile, we came to a small descent, which we descended by means of a ladder. We then walked, almost upon our toes, along a ledge, overlooking a kind of precipice, and then descended another ladder, and so, continuing our descent, until we reached the 'lode' which the miners were working.
>
> It is extremely dangerous descending a copper mine of this description, as the least false step would throw you over the precipice and dash you to atoms. My companion told me that he once took the Master of a vessel down this mine, and that this person was so terrified and became so nervous that he was obliged to have two miners to help him up again!
>
> We were all provided with candles and the light flashed upon the glittering ore and half naked figures of the dusky skinned men in a most picturesque manner."[3]

3. Vivian A. Webber, *Journal of a Voyage Round Cape Horn, 1859.*

There were railways after 1850, but most of the ore was carried to the coast by trains of mules and donkeys, each with a standard load of two hundred weight. At the head of each train was usually an old mare, the 'Madrina', from whose neck a bell was hung, and the side of every track was littered with the carcasses of animals that had died of exhaustion, hunger and thirst.

In Valparaiso the El Puerto, the port, lay at the foot of the cliffs which ran up behind the town, and on the top of these, reached by rough paths, were the dance-halls and drinking dives for which the area was famous – all frequented by seamen in Victorian times.

Valparaiso, circa 1834 (probably sketched by one of Darwin's companions on the 'Beagle').

Of these the most famous, or rather infamous, were the three 'Tops', Fore-Top, Main Top and Mizzen Top. They were little more than brothels and the women who plied their trade there frequently passed on venereal diseases, as well as emptying their clients pockets as soon as they could get them drunk enough.

In 1873 a severe earthquake struck Valparaiso and part of the cliffs including the Fore, Main and Mizzen Tops were destroyed, along with their barmen, dance-hall girls and customers. Down below in El Puerto itself, the drinking dens and brothels were all concentrated on a notorious thoroughfare which they used to call Gaff Tops'l Street. One of the lanes running up the hill called Calle Clave constituted the red-light district, lined with ladies of the night sitting on their doorsteps and inviting passers-by to come in and have a glass of wine. Drinks were of the chain lightning variety, mostly red and white vino, but also lager beer, and every kind of rotgut liquor. One of the most potent was a mixture called Aguadiente, which was purveyed in sausage shaped goat-skins, and was very cheap.

There was always a shifting population of unemployed seamen, mostly deserters from ships, who slept rough, beach combed, or got jobs as chuckers-out in dance-halls. When they became too much of a nuisance, the police would arrest them; they could languish in jail for weeks; often months. Sometimes the British Consul would help, but their best chance was always to get a job on a ship through the intermediary of a boarding house master, to whom they would have to mortgage their wages for several months in advance.

Corruption among port officials in ports up and down the west coast was notorious. An interesting example is quoted by Captain John Roose of the barque *Theta* while in Carrizal Bajo in April 1875 in a letter to the *Cambrian* (11th June 1875):

> "I entered the port of Carrizal on the tenth of March last, and was visited in the usual form by a Custom-House officer. I offered him my list of cargo and stores. He told me it was of no importance – that he would call for it some other day. On the 14th of the same month my steward was on shore and got into a midnight brawl with some natives, and was taken to prison. He was tried next day and fined seven dollars and fifty cents. The fine was paid, and he came on board in a dreadful state. I saw he would be of no service to me or the ship for some time. When I told him to leave my cabin he struck me twice with his fist (I at the same time as weak as a child through illness). I sent for the Captain of the Port to take him out of the ship, and he complied with my request. The next day the steward was tried for the assault. The result I do not know, and I have not seen him since.
>
> On the same day I was astonished to be so politely saluted by the Minister of Customs, soliciting my favours, as he, the Captain of Port, and others had formed a company and established a butcher's shop in the port. I was sorry I did not know in time; but when I made enquiries

I found that to purchase beef from these officials, etc., it would cost two cents per pound more than I was paying Lazaro for beef. I of course declined the honour of getting my beef from such a respectable firm. On the 15th my steward gave information that I had a lot of contraband goods on board. The Captain of the Port was perfectly justified in searching the ship, and he came on board on the 24th with the officers, my late steward, and a well-known sailor's crimp as interpreter. The Captain commenced by ordering certain sections of the cabin to be taken out and the bulkheads to be knocked down, saying that he would respond for all damages. Then he ordered the steward to go in my state room to select my new shirts from the old; and eventually he selected 22 linen shirts that had not been washed. It seems that removing goods that have not been washed is lawful. (!)

At this time I was on shore sick, and these proceedings were carried on during my absence. About 5 o'clock I was sent for. I went on board and found Don Valentine as jolly as a lord, with a bottle of whiskey and some cheese and sweet biscuits before him. I soon found that he knew how to enjoy himself. He had ordered the mate to supply him with these things. I was not inclined to be less polite than my chief. When Don Valentine had finished one bottle I ordered another to be opened. He was then in high glee. He ordered his officers to pack up the goods seized and take them on shore. He stopped; and I give you my word of honour that he did justice to the second bottle of whiskey. Well, this got into the poor old Don's legs, and they refused duty; and, weak as I was, I had to lead and carry the good Don to his home, where I seated him in a chair, and in one minute after he was fast asleep. Before this the good Don had told me and my interpreter (Mr. Holloway) that the new butcher's establishment belonged to him alone; and that if any English ship came here and did not deal with him he would humbug them the whole time, or at every opportunity; also, if I had taken my beef from him this would not have happened, viz: the seizure of my goods. I have been trading to Chile for the last twelve years, but never heard so much obscene language from one man's mouth in my life as I heard from the Captain of the Port of Carrizal on the evening of the 21st of March last."

Iquique, further up the coast, was the main nitrate port, and after Valparaiso, the most notorious. It was not uncommon for a hundred sailing ships to be lying at anchor in its bay, stretching from headland to headland. Ashore, there was a huge changing population of seamen, deserters, and beachcombers, all scrounging a living as best they could, and the usual collection of boarding houses, drinking dens and dance halls.

Iquique, circa 1903.
(National Photo Agency/National Maritime Museum, Greenwich)

Callao came a close third to Valparaiso and Iquique, and tended to be a regular port of call for homeward bound vessels. It was the clearing house for the Guano carriers, and vessels often topped up there with hides, bones (for fertiliser) and coffee.

Callao was particularly notorious for its boarding house crimps, and had the usual floating population of deserters, out-of-work seamen, beachcombers and drifters of every kind.

The names of some of the bars deserve recording, such as 'Smokey Joe's', 'The Neptune', 'The Cape Horn' and the 'Live and Let Live'. Of Coquimbo, a Swansea gentlemen, Vivian A. Webber of Sketty, in his *Journal of a Voyage to Cape Horn*, published in 1859, wrote of the port of La Serena: 'High and low conspire to clean out any unwary foreigner or stranger within their gates. I do not think that I ever met with such an unblushing set of rogues.'

The crews of the Swansea barques were no strangers to Chilean or Peruvian jails; at the height of the copper ore trade it was unusual for the ships not to report at least one deserter. A common entry would be along the lines of: 'Deserted, taking their belongings with them, at an unknown hour of the night'. If the vessel was lying at an anchorage the deserters would take the ship's boat, usually beaching it at some point where it could be recovered by the ship.

In due course a boarding house master would find them a berth on another vessel, short-handed through sickness or desertion, as usual demanding an advance of the man's wages from the Captain of the ship, who would also be expected to pay a fee to the boarding house master. Not all such seamen were deliberate deserters; many of them had been given a 'Mickey Finn' (a drugged drink), robbed, and just missed their ship.

The Swansea men who visited the west coast of South America could never forget it, and talked of towns ankle-deep in dust, swarming with flies and narrow dirty streets filled with drinking dens and dance-halls. There was always the same background of steep hills and cliffs, without trees, and simmering in the heat. And off them lay the heaving anchorages where their ships laid at anchor lifting and falling to the perpetual Pacific swell.

Of the public houses on the Strand at Swansea they returned to, only the 'Cardiff Arms' remains, passed every day by shoppers on their way to Sainsbury's and Homebase, over ground where the barques once floated. It cannot be long before it is pulled down and who will then think of the tales that must have been told and the songs sung over its polished mahogany bar? The ghosts of the old seaman will by then have truly all departed.

THE GUANO AND NITRATE TRADES OF CHILE AND PERU

Peru lies between Chile in the south and Bolivia to the north. It extends much further inland, but as in Chile, the coastal area is still separated off from the rest of the country by the Andes range of mountains. Even today, much of the interior is largely unexplored and forms part of the Amazon rain forest, to which it is connected by the great river Maranon, a tributary of the Amazon.

Like Chile, the population is Spanish-speaking and a mixture of Indian and Spanish descent, with the rural population predominantly Indian. As in the 19th century, its chief exports are still mining products: iron ore, copper, zinc, lead and silver, together with some nitrates. In terms of area, it is the fourth largest country in South America. Again as with Chile, its capital Lima has its satellite port of Callao, only a few miles distant.

Coal, copper ore, guano, nitrates and wheat were the main bulk cargoes carried by the sailing ships, and Swansea vessels were very much involved in the nitrate and guano trades. Guano consists of the droppings of countless seabirds which have accumulated on the volcanic islands off the west coast of South America, mainly in Peru and Bolivia. The deposits, which grow to resemble white cliffs, could be as much as 100 feet deep, and form probably the best fertiliser in the world. The best quality of guano included the droppings of the seals which frequented the rocks in vast numbers. The chief sources were Forsyth Island, Constitution Roads, Mexilones Bay and Paquiqui in Bolivia, and the Chincha Islands of Peru.

A pilot shipment, consigned by the Peruvian government, arrived in Liverpool in July 1839, and after its properties as a fertiliser had been recognised and acclaimed by agricultural writers, guano rapidly became sought after all over the world. The guano shipments were controlled by British agents appointed by the Peruvian

Loading bridge, Antofagasta.

government, and every vessel had to call at Callao, the port of the capital Lima, to be surveyed before being chartered.

It was an extremely nasty cargo to carry, since it gave off dangerous ammonia fumes when wet, which could sometimes even affect the stability of the ship, and could corrode ironwork in the hold, as well as rotting sails and rigging. When loading, ships lay under chutes, wreathed in clouds of noxious yellow dust. But at least there was the consolation that when the cargo was loaded and the hatches battened down it could be washed off without too much difficulty.

The nitrate ports were spread along the coast and nearly all had similar characteristics: an open anchorage, with a cliff face towering up behind a small town. Iquique was the largest, but there were many of them, such as Talcahuano, Taltal, Antofagasta, Tocopilla, and small ones like Caleta Buena. Inland sun-scorched desert stretched to the foothills of the Andes mountains, crowned with snow. Most of the anchorages were treacherous places with dangerous currents and difficult anchorages. Talcahuano was a happy exception, sheltered by the island of Quiriguina in Conception Bay.

From the beginnings of the trade in the 1830s it grew rapidly and the coast in late Victorian times was filled with craft of all sizes, but particularly the big nitrate clippers, for which it became famous. The reason is not to hard to understand, because nitrate was used for making sulphuric and nitric acid, important in various chemical processes, the chief of which was making saltpetre, used for the manufacture of gunpowder. It was also valuable in its original state as a top-dressing in farming operations.

It was dug out of the ground and put into bags which were delivered into the ship's boats, which after they had received say fifty bags were rowed out to the ships, perhaps a quarter or half a mile away.

It was an extremely dangerous cargo to carry, especially as the bags tended to rot. It was highly soluble in water and gives off deadly fumes when it evaporates. But the greatest danger is from fire; which could only be put out by spraying with water which itself contains nitrate. When vessels were loading nitrate it was usual to keep barrels of nitrate water in readiness by each hatch.

The cargo came out in barges and the bags of nitrate were lifted up, one at a time by the crew, using a hand-winch. The method of loading was as interesting as it was peculiar; a square of bags was built up on the floor of the hold, directly below the hatch square and one solitary local stevedore would place each bag in the position where it had to stay. This one man would load the whole ship. In order to distribute the weight to avoid excessive rolling, the bags were built up in a pyramid both in the lower hold and 'tween decks. It seems extraordinary that cargoes loaded in such a manner never shifted, but contemporary accounts all seem to agree. Apparently the nitrate, on the long voyage home, would bind into a solid mass, which then had to be dug out at the discharging port with a pick and spade.

Any account of the guano and nitrate trades would be incomplete without mentioning the horrific conditions under which Chinese labour was recruited to the coast to work these unpleasant substances.

Chinese colonies were attracted to the coast by an unholy alliance between labour crimps in Macao and along the South China coast.

They were deluded into signing contracts, hoping for a better life in the New World, but many died on the long voyage from China. The more fortunate ones were put to work on the sugar plantations, but the majority were sent to work on the horrific guano islands such as Forsyth Island or the Chinca Islands, or to dig out nitrates.

On the guano islands they were forced to dig up to six tons of guano a day, seven days a week. Expectation of life was short and suicides were common, sometimes committed by jumping into the loading chutes.

In 1855 the British parliament introduced severe restrictions on the carriage of coolies on British ships, followed a few years later by the USA. But the trade still continued illicitly, and only came to an end when owing to a fall in the supplies of guano, the Chilean and Peruvian governments later restricted exports of guano, reserving the larger part for their own agriculture.

SAN FRANCISCO

Until the 1840s, the Swansea copper barques were the most frequent sailing vessels to double Cape Horn, except for a small number of American vessels trading to the Spanish-speaking settlements along the Californian coast, and whaling ships from the New England ports.

San Francisco, 1855. More than half the vessels had been abandoned by their crews in the Californian gold rush.

The situation changed dramatically when the USA declared war on Mexico in the mid-1840s. By 1848 Mexico City had been occupied and California ceded to the United States. At the same time, by an amazing coincidence, gold was discovered, and the celebrated California gold rush of 1849 began. It was the sailing ships rounding Cape Horn that opened up San Francisco, and provided the main link between the east and west coasts before the transcontinental railroad was completed in 1869. And after its completion, San Francisco continued to be one of the world's busiest sailing ship ports, stimulated chiefly by the grain trade, which continued to offer commercially viable cargoes until well after the end of the 19th century.

California was first opened up by the Spanish in the 16th century, seeking to establish a trade route between the Philippines and Mexico. The entrance to San Francisco's huge land locked bay eluded the first explorers, until it was discovered by accident in 1769. From that point on San Francisco as a port began to grow, with increasing numbers of American ships engaged in whaling or the fur-trade or to exchange China goods for hides and tallow. The combined effects of California becoming a state of the USA and the gold rush of 1848 created an explosion of activity and prosperity.

It can truly be said that California, and San Francisco in particular, was opened up by gold, wheat, and the sailing ships. The settlers who drove westward along the Oregon trail, and later the transcontinental railroad, started the process, but it was those who came by sea that began the immigrant explosion.

At the peak of the wheat trade San Francisco alone loaded five hundred grain ships in one season, and Portland another hundred. It was the thrusting energy of

San Francisco entrepreneurs that generated the trade in the first place; food was needed for the hundreds of thousands of immigrants who poured in, attracted by gold and at the prospect of life in the New World.

It was not long before there was a surplus, and the famous wheat trade of San Francisco had commenced. In about 1860 almost half the grain exports went to the UK, with Australia taking a quarter and the rest mainly to New York and China.

The California coast, prone to dense fogs rolling off the Pacific, with its rocky stones and fast currents, was a dangerous place for sailing ships, and hundreds came to grief through stranding, collision, capsizing or foundering. But it is not just bad visibility and treacherous currents, there are winter storms that blow in with the southerly gales. Reefs abound, and once a vessel went ashore it would soon be pounded to pieces. And lying across the entrance to the Golden Gate there was the hazard of a huge crescent-shaped bar.

The Barbary Coast, as it came to be known, soon became a name to conjure with. Above all it was a cosmopolitan port filled with gold-seekers, adventurers, miners, gamblers and deserters, all hoping to make their fortune. The successful and well-to-do citizens were already building themselves fine houses on the hill; there was an opera house, and fashions from New York and Paris were all the rage. Down below, along the waterfront, it was a sort of frontier town, rather like a scene from the James Stewart movie, 'Destry Rides Again', constructed of frame houses, lean-tos and tents.

Saloons, dance-halls and brothels abounded, strung along Pacific Street, which led up from the waterfront; it was this street of vice that came to be called the 'Barbary Coast'.

Sailors' taverns, whether they were pubs, saloons, ale-houses, or gin-palaces always lined the dock area in every port, and San Francisco was no exception. Catering largely for a British, American, or North European clientele, the signs outside them were almost universal, with sexy mermaids, dolphins, anchors, busty ladies and negro heads predominating. The names tended to be the same everywhere, like 'First and Last', 'Smokey Joes', 'Pacific', 'Atlantic', 'Chain Locker', 'The Look-out' and the 'Cape Horn Inn'.

The interiors ranged from sawdust floors (replete with spittoons) which were hosed out every morning, with wooden benches and settees, to full scale Victoriana, with velvet curtains, plush seats, mirrors everywhere, ornate lamps, and always a honky-tonk piano. The waitresses usually doubled as ladies of the night, flattering their victims into a state of drunkenness where they could be rolled for their money and then turfed into the street, or drugged and shanghaied.

Some had extraordinary signs above their doorways, such as Madame Lucy's 'Ye Olde Whore Shoppe' in Sacramento Street, or 'Madame Lazerene's' which displayed a huge red iron rooster above the gilded name of the house: 'Sign of the Red Rooster'.

Shanghaiing; a doped seaman is hauled on board.
(Courtesy Bill Pickelhaupt, San Francisco)

Others that lived on in the memories of Swansea seamen who had visited them, were 'Horsine Brown's', 'Hells Kitchen', 'Paddy Ryan's' and the ' Cowboy Maggie', reigned over by Cowboy Maggie, popularly designated as the 'Queen of the Barbary Coast'.

Things were not much different in the dance-halls, where the hostesses were of the same ilk. A famous one was a large lady known as the Galloping Cow, and the story goes that she once broke the back of a man who ungallantly tried to steal a kiss.

Another was 'Lady Jane Grey', popularly believed to be the offspring of an English Earl and a dance-hostess. She souped up her act by wearing a cardboard tiara decorated with tinsel and eventually retired from the dance-floor to become a pick-pocket.

John 'Shanghai Chicken' Devine made his career in bare-knuckle fighting and shanghaiing. A hard-nosed criminal, he was arrested eighty-eight times in less than ten years, and was hanged in 1873.
(Courtesy Bill Pickelhaupt, San Francisco)

The other group of individuals for which San Francisco earned an unsavoury reputation were the boarding-house masters and crimps, who specialised in doping seamen and shanghaiing them aboard ships in return for an advance on their wages. There was always a floating population of sailormen in most ports: deserters, or men who had missed their ship, or signed off because of illness or injury and these naturally gravitated to the seamen's boarding houses. It was all part of an evil though recognised system, in which the captains of ships, willingly or unwillingly played their part.

One well-known lady crimp was known as Chloroform Kate, who was assisted by two henchmen, called Harry the Ape and Red Simms. When Kate died, Harry arranged an impressive, if somewhat unusual funeral. She was wrapped up in an old tarpaulin, with a fly-blown picture of the famous American clipper, *Sovereign of the Seas*, placed upon her breast, and then put aboard the San Francisco 'garbage barge', to be towed out to sea and consigned to the deep with the rest of its contents.

A particularly notorious crimp was a red-bearded Irishman known as 'Shanghai Kelly'. His boarding-house jutted out over the tide water and was equipped with three trap-doors, through which his unconscious wares could be lowered into boats and rowed out to the waiting ship.

The boarding-house masters, whose names alone speak volumes, like Bully Williams, Shanghai Kelly, Three-Finger Daly and Scabby Joe, were usually hand in glove with crooked local politicians. They were constantly opposed by men like the Rev. J. Fell or 'Fighting Karney' who ran the Seamen's Missions. Some of the padres had been to sea themselves, and were able to take care of themselves. They would arrange boxing matches and races between ship's boats, but most of all they looked after the ship's apprentices, organising shore expeditions and dances where they could meet respectable girls. The charity and kindness of such men lived on in the memory of many a Swansea seaman, reminiscing in old age about his days in the Barbary Coast.

One notorious American sailing ship captain known as 'Bully Waterman' became religious in his old age, and began boarding ships in San Francisco harbour in order to preach the gospel. He was recognised by old seamen who had experienced his brutality, and thrown overboard by the indignant crew, only narrowly escaping drowning by the timely arrival of a police water patrol.

The struggle between the Shanghaiers and the local authorities went on all through the sailing ship era. As long as there was money to be made by crimping a seaman, the trade was bound to continue. It diminished along with the sailing ship and by the beginning of World War One it had become a thing of the past.

12.

James Harris Jnr., Cape Horner and Marine Painter

ONE PARTICULAR SWANSEA Cape Horner deserves a chapter to himself, for he, together with his father James Staden Harris, were marine painters of the highest quality, whose work is now beginning to receive the international recognition it has long deserved. James Staden Harris was born in Exeter in 1810. His father, John Harris, with his wife, two sons and a daughter, moved to Swansea in 1828, where he set up in business in Wind Street as a carver and gilder of picture frames. When he died in 1836, the family carried on the business and an advertisement in the *Cambrian*, 1st October 1842, lists it under the name of James Harris at 40, Wind Street.

In August 1846 his mother married again, and it is thought James' interest in the business ceased at about that time. By 1850 James is listed as an artist living in Mumbles. In 1869 Pearce Brown's trade directory gives his address as Underhill Cottage, Oystermouth, which now cannot be identified. In 1882 he moved to Bryn House, Reynoldston, Gower, where he died aged 77 on 19th June 1887.

His son, James Harris Junior (1847-1925), is increasingly becoming internationally recognised as one of the leading marine artists of his day, but even so it is generally agreed that James Harris Senior was even better.

Annie Harris, the sister of James Harris Junior, married George Standfield Walters (1837-1924), the well-known Liverpool marine painter, who studied under the celebrated London artist Edward Duncan (1803-1883). Duncan was a frequent visitor to Swansea and Gower, where between 1845 and 1882 he did many of his best paintings. There is good evidence to suggest that during these visits James Harris Junior was his pupil.

James Harris Junior left nine sketch-books of which six remain safely intact; the others, regrettably, have almost certainly been broken up.

In 1867 he was commissioned by C. R. M. Talbot to do a return voyage round Cape Horn, sketching and painting all that he could see, both on shipboard, of coastal scenes, and ashore.

C. R. M. Talbot regularly travelled to various ports in the Mediterranean in one or other of the different yachts that he owned. There is no record that James Harris

Photograph is believed to be of James Harris Jnr., Cape Horner and marine painter (1847-1925).

accompanied him on any of these journeys, though the Rev. Calvert Jones, the well-known marine painter and photographer (he was a friend of Henry Fox Talbot), frequently did so.

Calvert Jones was more within Talbot's social class, and it has been suggested that the painting 'Shipping off Messina' by James Harris, at present in the Glynn Vivian Art Gallery Collection in Swansea, might have been based on sketches made by Calvert Jones. One of the sketch-books contains pictures, mostly in pencil, made during a visit to Maritzburg in Natal in 1882.

The portrait on page 199 is believed to be that of James Harris Junior towards the end of his life, but its authenticity is not certain. The younger Harris, nearer to us in time than his father remains, for all his many paintings and sketches, a shadowy and elusive figure.

The seven sketch-books which have survived may be briefly described as follows:

(1) *Undated*, but the Battery on Mumbles Head indicates after 1860. Mostly pencil sketches of local boats.
(2) *April 1st, 1866*. Pencil sketches of local boats and occasional landscape in watercolour.
(3) *June 1867*. 'Dordrecht'. June 1867. Watercolour sketches of canals, windmills and buildings with sailing barges.
(4) *September 29th, 1867*. The most interesting. Sketches of coastlines, details of rigging, etc. Describes a voyage from the Clyde, round Cape Horn. This sketch-book relates to a number of finished watercolours which are signed and dated.
(5) *July 5th, 1871*. Pencil sketches of local boats and figures. Included is a drawing of an artist at work.
(6) *October 2nd, 1880*. Sketches made on a voyage – Capetown on October 25th 1880, Ascension Island, Constantinople April 21st 1881. The pencil notes are difficult to read.
(7) *May 1882*. Maritzburg, Natal. Mostly pencil drawings of trees and vegetation.

It is known that between 1902-10 he lived at Reynoldston, the home of his mother who died in 1890. In 1911 he was living at 1, Castleton Place, Mumbles, and then in 1912 he was living at 8, Southwood Road, Ramsgate, which could have been the home of his sister Annie, who married George Stanfield Waters. In 1921 he returned to Mumbles, where he lodged at Seacot, Limekiln Road, the home of the Williams family. It is said that Mrs. Williams paid for his burial and grave. He died on December 11th 1925, aged 72 years, was buried in Oystermouth Cemetery (Grave No. 4515). His epitaph was 'in the evening, it shall be light'.

Note: I am indebted to Dr Rodney Howell of Swansea for permission to quote from his researches into the Harris's, father and son, and his text for 'Under Sail' published in 1987 by Swansea Museum Services. (See Obituary, *South Wales Evening Post*, 12/12/1925.)

13.

The Swansea Pilot Cutters

NO WORK DEALING WITH Swansea's copper barques and Cape Horners would be complete without at least a brief description of Swansea's distinctive pilot cutters.

Writing in the early 1920s, W. H. Jones, in his *History of the Port of Swansea*, mentioned that there was still at work in Swansea a pilot named Elijah Williams who had 'sixty-seven years experience of the sea, including piloting from Swansea since 1873, 19 voyages round the Horn, and 13 to Cuba in the ore trade'. During Elijah Williams' early days as a pilot each man worked for himself, and the most enterprising of them would not just wait for a ship off Mumbles Head, where the vessels were entitled to look for pilots, but sail down the Bristol Channel to Lundy and beyond to pick up incoming vessels. The competition was considerable, and naturally the most adventurous and skilful pilots picked up the cream of the work.

The Bristol pilots, through the Bristol Corporation, claimed pilotage jurisdiction over the greater part of the Bristol Channel, but this was never accepted by Swansea by virtue of the Swansea Harbour Act of 1791 (31 George III C. 83). The pilot cutters had to be fast, seaworthy, and easy to handle. The Mumbles Roads, four miles across the bay, did not provide very good holding ground in bad weather and the services of pilots was often urgently needed.

When putting a pilot aboard a ship it was essential to have a sail rig that could be easily handled while alongside and hoisted again quickly when getting away. The rig that lent itself most readily to these requirements was known as a 'Shallop' rig and consisted of a main and foremast, former raked markedly backwards, with very short gaffs. The main boom overhung the stern by about half its length. The rake of the main mast was designed to counteract the effect of the very short gaffs, which made the rig so much easier to handle. A typical example is the *Grenfell*, built in Swansea in 1865 by Philip Bevan, and which is illustrated on page 202. This drawing, which was a reconstruction from a Swansea sail-maker's notebook, and the details were filled in from old photographs and the descriptions of pilots who sailed in them.

The following list of Swansea pilot boats in service extracted from the *Mariners Mirror*, 1944, page 122, by Mr. J. F. Coates, the tonnages and dimensions being provided by Mr. A. G. Moffat of Swansea.

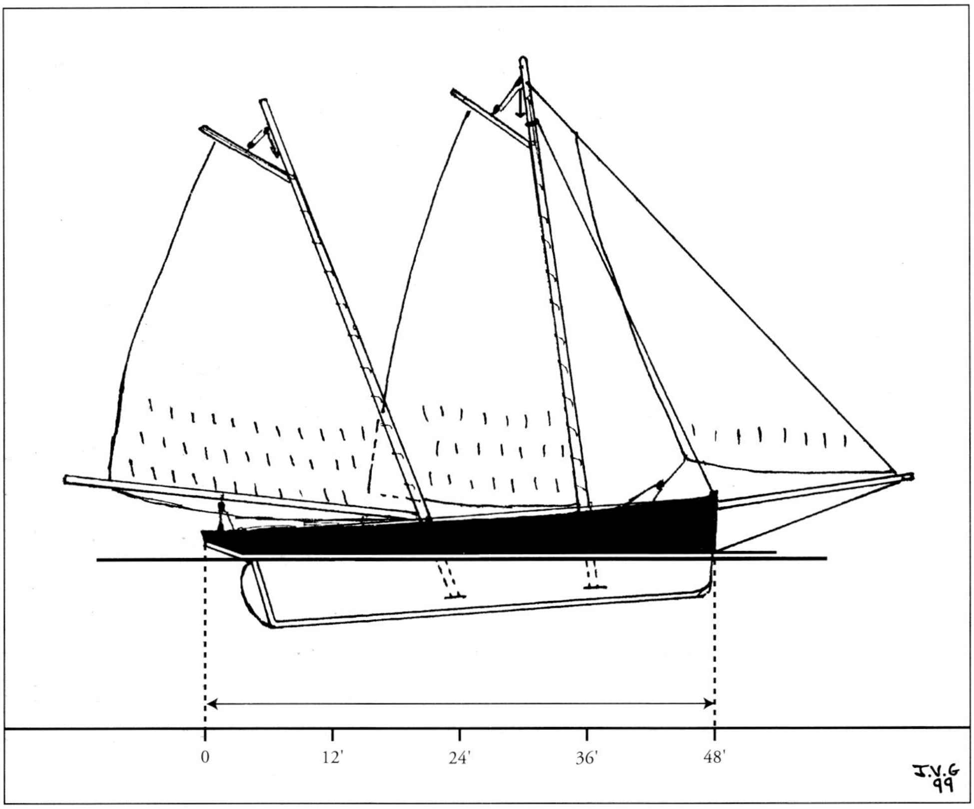

Sketch plan of a Swansea pilot boat, circa, 1870.

The Swansea pilot boat 'Vivian', 1893.
('Mariner's Mirror', July 1944.)

PILOT BOATS IN SERVICE, c.1840-60

No. 1	*Singleton*	No. 2	*Tom Rosser*	No. 3	*Zion*	No. 4	*Faith*
No. 5	*Swanzey*	No. 6	*Sarah*	No. 7	*Henry*	No. 8	*Neptune*
No. 9	*Henry*	No. 10	*Providence*	No. 11	*Vivian*		

DECKED BOATS IN SERVICE UNTIL ABOUT 1895

No.	*Name*	*Owner*	*Built*	*By*	*At*	*Tons*	*Length ft.*	*Breadth ft.*
1	*Vivian*	D. Owen	1863	Bevan	Swansea	22.39	45.6	11.3
2	*Vigilant*	W. Rosser	1859	Bowen	Swansea	17.12	41.6	10.9
3	*Rival*	J. Rosser	1861	Bevan	Swansea	21.15	43.0	11.6
4	*Benson*	J. Rosser	1870	Bowen	Swansea	28.27	51.0	13.6
5	*Lizzie*	–	1875	Bowen	Swansea (or Cocks, Bideford)	33.13	52.0	13.0
6	*Glance*	W. Rosser	1863	Perkins	Bideford	21.77	46.0	11.6
7	*Alarm*	–	1860	Bowen	Swansea	19.43	42.0	11.0
8	*F.W.F.*	G. Fox	1863	Bowen	Swansea (or Perkins, Bideford)	26.16	48.0	11.6
9	*Grenfell*	Burton	1865	Bevan	Swansea	29.29	50.0	13.0
10	*Camilla*	T. Brown	1960	Bowen	Swansea	18.70	45.6	10.0
11	*Charles Bath*	Bibby	1865	Cocks	Bideford	19.00	49.6	12.0

14.

The Shantymen

MORE THAN ONE OLD Cape Horner, reminiscing about his days on the west coast of South America, describes how ships used to lie at anchor in the bays of ports strung along the coast, and sometimes in the evening, when the work of loading and unloading was done, the sounds of some haunting chorus would come floating across the sea from the Welsh copper barques at anchor there.

It recalls the writer's own feelings while at sea during the Second World War when the old Light Programme, which could be heard halfway across the Atlantic, and as far as Sicily in the Mediterranean, transmitted a music programme for the forces, which always finished with a rich Welsh voice introducing the final song, which was 'We'll keep a welcome in the valleys'. In those days Wales did not have the same meaning for me as it does today, but even then, I think the effect was still the same.

Chamber's dictionary defines 'Shanty' as a 'song with a chorus, sung by sailors while heaving at a capstan, or the like', and that is probably as good a definition of its history and origin as one can get. It used to be said that the singing of a shanty in chorus was as good as three extra men in a gang heaving at a capstan or on a sheet, and on Welsh vessels where singing was in their bones and blood, it was especially important.

There was usually one seaman, probably an older hand, who had a good memory and a powerful voice, who naturally took the lead, and he would be the 'shantyman'. It was not a specific job that carried extra pay, but good shantymen were essential to the easy running of the ship, and as such highly regarded by the captain, and consequently were never short of a berth. Ideally, there would be two shantymen, one for each of the two watches into which the crew were divided.

There were various kinds of shanties, with different metres, appropriate to different tasks such as heaving up the anchor on the capstan, hauling on ropes to trim the sails, or the long hours spent working at the pumps on wooden ships, which always leaked. A lot has been written about obscenity in the shanties, and of course sailing ship seamen were by no means saints. But mostly they were decent men, who could be roused, and their language might properly be described as more

'Elizabethan' than lewd. Joanna C. Colcord in her *Songs of American Sailormen* sums it up as a 'jovial, forthright almost wholesome obscenity'.

Quite a lot of the early influence on the songs of the shantymen sung probably derived from the Negroes who were found in such numbers on ships as seamen and cooks. They came from the West Indies and the southern United States and brought their nostalgic plantation songs with them, which had been adapted to the labour songs of the black stevedores and wharf workers in the southern American ports.

Whole books have been filled with songs the shantymen sang, and who originally composed the words or even what they were can never be definitely stated; they just naturally grew up and became a tradition akin to folk-lore. The following verses from some of the best known give some idea of their flavour:

WHISKY FOR MY JOHNNY

Then up aloft that yard must go,
Whisky for my Johnny
Oh, whisky is the life of man,
Whisky, Johnny, whisky.

I thought I heard the old man say,
Whisky for my Johnny
We are bound away this very day,
Whisky, Johnny, whisky.

A dollar a day is a white man's pay
Whisky for my Johnny,
Oh whisky killed my sister, Sue,
Whisky, Johnny, whisky.

Whisky's gone, what shall I do?
Whisky, Johnny, whisky.
Oh, whisky's gone, and I'll go too,
Whisky, Johnny, whisky.

LEAVE HER JOHNNY, LEAVE HER

Oh, a dollar a day is a white man's pay,
Leave her, Johnny, leave her,
To pump all night and work all day,
Leave her, Johnny, leave her.

The Old Man shouts, the pumps stand by,
    Leave her, Johnny, leave her
Oh, we can never suck her dry.
    Leave her, Johnny, leave her.

It's pump or drown the Old Man said
Or else damn soon you'll all be dead
The rats have gone on' we, the crew
It's time be damned that we went too,
    Leave her, Johnny, leave her.

A LONG TIME AGO

In the Blackball Line I served my time
That's the line where I wasted my prime
There's tinkers and tailors, and fakirs an' all,
They've all shipped as ABs aboard the *Blackball.*
It's for tops'l halyards, the Mate he will roar,
It's lay along Paddy, you son of a whore!

And finally there was 'Old Swansea Town', the words which seem to summarise all that I have tried to portray in this book.

Good-bye my lovely Nancy,
Ten thousand times adieu;
I'm going away for to leave you,
Once more to part from you,
Once more to part from you, fine girls,
You are the girls we adore,
But still we hope to see,
Old Swansea Town once more, fine girls.

Now we are out on the ocean,
And you are far behind,
Kind letters we will write to you,
With the secrets of our minds,
With the secrets of our minds, fine girls,
You are the girls we adore,
But still we hope to see,
Old Swansea Town once more, fine girls.

There is a dreadful storm arising,
We can see it coming on;
The night is dark and stormy,
And our road seems all forlorn,
Our good ship is tossed about,
Our rigging is all torn;
But still we live in hope to see,
Old Swansea Town once more, fine girls.

Now our storms are over,
And we are safe on shore;
We will drink success
To all the girls,
And the one that I adore.
We will drink strong wine and brandy, too
And we will make those taverns roar;
And When our money is spent and gone,
We'll go around Cape Horn for more, fine girls.

Chorus:
Old Swansea Town once more,
And still we hope to see,
Old Swansea Town once more.

15.

Women at Sea

THE PRESENCE OF WOMEN AT SEA falls into categories and periods. There were those who travelled as passengers, and one thinks of those wives who accompanied their husbands to various parts of the British Empire in colonial days. Most of all perhaps, to India, on the long passage on an East Indianmen via the Cape of Good Hope.

Then there were the wives of emigrants, seeking a new life in the colonies of Australia, Tasmania and New Zealand. And there were another kind, the heart-rending women transported to such places as Botany Bay for some minor offence, often committed because of desperate poverty and the need to feed their family, or avoid the horrors of the workhouse. It is a relief to learn that a large proportion of these served out their sentence, married and raised families, and some even became the owners of prosperous businesses.

On merchant ships they were almost invariably the family of the captain. The captain's wife could range from being the friend of the crew, especially the apprentices, mending socks and tending sick cases, to selfish nagging miserly harpies who sometimes came near to provoking a mutiny on a ship.

The captain of a ship was lord of all he surveyed. If a seaman wished to speak to him he would first remove his hat and place it under his arm, before asking permission to speak.

A scheming woman nagging a husband with such powers could make life on board a misery for everyone. Some encouraged their husbands to persecute the crew with the object of causing them to desert on reaching a foreign port, in which case their outstanding wages would be forfeit and go straight into the captain's pocket. Cutting down on the crew's food and buying the cheapest provisions available in foreign ports was another source of profit for an unscrupulous shipmaster, and sometimes such women got into their hands the doling out of supplies on board.

It is a relief to turn to a wife who was totally kind, Mrs. Powles, who sailed on her husband's ship the *John O'Gaunt*. Her only son, Thomas Mariner Powles, was born on that ship, and she sailed with him from their marriage in 1887 until sadly, she died twenty years later. Both the Powles were great musicians and carried an

'Herbert', 1867 (flying Norwegian flag and Burgess House flag).
(Private Collection)

The barque 'Eta', by James Harris Jnr.
(Private Collection)

The 'Vanduara'.
(Private Collection)

organ and piano to sea with them. Seamen sailing with them for the first time were astounded to be invited into the saloon for a concert, and everyone, high or low, was treated with unfailing courtesy, liberality, and kindness (Basil Lubbock, *Last of the Windjammers*, Vol. I).

Wives at sea were often skilled navigators, and there was more than one case where, due to accident or sickness, or disaster on board, the captain's wife navigated the ship safely into port, and sometimes even taking command.

In the little sea-faring communities that ringed the British Isles, so often a family depended for its happiness, health and prosperity on a woman who held it together in good times and bad. Such a one was Annie Harding, born in Devon in 1865, the daughter of the master of an old smack called the *Dahlia*, which traded up and down the Bristol Channel.

From an early age she went to sea with her father, and as soon as she left school sailed with him as one of his crew. Her son, Captain J. W. Slade of Appledore, who raised himself by hard work to Master and eventually a shipowner, and author of *Out of Appledore* and *West Country Trading Ketches*, wrote of her as follows:

> "She was taught the compass and how to steer by it, how to trim the sails by the wind or going off the wind, but better still she knew the land and sets of the tides according to the time it ebbed or flowed. She became a good judge of weather conditions. In fact she knew enough to be left in charge of the watch while her father slept.
>
> In those days, most of the old ships were veritable death traps. When at sea they were generally kept floating by continual pumping. The first stop after leaving Lydney (with coal for North Devon) was Penarth mud. There the crew would rest from the pump because the mud would get in the leaks to stop the water from flowing in. Often I have heard mother talk of the hours she stood at the pump in her high laced boots with hardly any time to rest or make a cup of tea till she eventually arrived at the port of discharge. Then all her spare time was taken up cooking and feeding the crew including herself."

When her father became ill she had to be the family's main breadwinner, working in a collar factory for about a £1 a week, which she gave to her mother. Eventually she married William K. Slade, who was mate of a schooner called the *Hawk*, built in Prince Edward Island, in the Maritime Provinces of Canada. He was illiterate but ambitious, and Annie taught him to read and write via the Bible and by exchanging letters to him, and correcting the ones he sent to her.

She was an excellent sea person and when he was given command of a ketch called the *Francis Beddoe* she went to sea with him and helped him acquire the art of navigation. She saved any spare money and as her son relates, it was the proudest day of her life when, her baby cradled under her arm, she went to the bank to pay in their first savings. Eventually they saved enough to buy a quarter share in a new schooner, named the *Alpha*. Later, as their financial position improved, Annie settled down ashore. They had enough money to buy a nice house, and all the time Annie made sure that money was saved. Another schooner, the *Ulellia*, was bought in partnership with a relative, George Quare. William Slade became managing owner and George Quare went to sea as Master, while Annie kept the books as well as tending to the house and her family.

After her death her son wrote of her: 'There were many fine women who were ideal sailor's wives in Appledore, but when I look back and think of the life of my mother, Rosina Annie Abigail Slade, I feel she was outstanding as a wife, as a helpmate, and a mother of whom we can all justifiably be proud.'

She died in November 1964, nearly reaching the age of a hundred, and keeping her faculties to the end. She refused to take the drugs her doctor ordered, saying, 'I know I'm on the verge and I want to see what it is like when I go'. She died, fully conscious, sensible, and unafraid.

Seamen tend to be superstitious folk and in the days of sail it was widely considered unlucky to have a woman on board. It is an odd thing, but looking through old records of disasters at sea, it is surprising how often it seems to be a captain's wife who dies or is drowned. Perhaps women were more physically vulnerable, but today that view is hardly current: almost the reverse, in fact. But as Basil Lubbock remarks, 'over and over again in the case of a shipwreck, she is lost.'

On one occasion two ships collided in bad visibility off Cape Horn, and were swept beam to beam together in the trough between two gigantic seas. The captain of one of them, fearing that his vessel was about to sink, attempted, just before the collision, to throw her on to the other ship. It did not succeed and she fell between the two and was crushed to death, the only victim of the accident.

On the 15th March 1910 the sailing ship *Swanhilda*, commanded by Captain Pine, sailed from Cardiff for the west coast of South America. On May 6th, having miscalculated her position in bad visibility, the vessel was carried by heavy seas and strong currents on to Cape St. Anthony. Recognising that his vessel was doomed, Captain Pine ordered the boats to be swung out, but before this could be done the vessel was swept broadside on to the rocky coast. A few members of the crew managed to make their way ashore, but all died of starvation save one, who was later found alive, but insane in a cave. He later recovered, and said that he had lost his reason after discovering the bodies of Captain Pine and his wife, still locked in each other's arms, washing about in the tide. The tragedy, as so often the case with ships going ashore, was that the hull remained more or less intact; had the crew remained on board until the weather calmed down they might all have survived.

In March 1872 the *Cambrian* newspaper reported the story of the barque *Lilly Palmer* from Swansea, commanded by Captain Fletcher of Liverpool. A watchman on Sable Island, on the north-east coast of Canada, heard a whining dog, and finding it, tied a lantern to its neck. It led him to a woman lying on the shore with a small child beside her, both suffering from severe exposure. The child died next day in hospital but the mother, described as 'refined and handsome', survived to tell her tale. *The Lilly Palmer*, bound for Halifax with merchandise and salt, had encountered a gale which lasted for five days, during which she lost her top foremast. Somehow they fitted a jury rig, but after another thirty days ran into a hurricane. The ship nearly lost all her foremast; the mizzen was split in two by a huge sea which made a hole in the ship's side and forced her on to her beam-end. Thus she drifted for several days and ended up 500 miles off course near Sable Island. She lost her rudder, and drifted on to a lee shore, where her bows stove in, she filled with water, and broke in two. The article does not say whether there were any other survivors, but the woman (presumably the captain's wife) was saved by a barking dog.

In Shanghai, China, on the 26th June 1904, a sad funeral took place at the quaintly named Bubbling Well Cemetery. It was of Susan Ellen Sainty, the wife of Captain Henry George Sainty, who died in childbirth in hospital in Shanghai while her husband's ship lay at anchor in the Yangste River. Her death occurred while her husband was on urgent ship's business, and he returned to find that he had lost both his wife and their child. Theirs had been a love match since their marriage in 1893, and she was 38 years old when she died. Family tradition has it that he was broken hearted, and never really recovered from her death, though he continued to command ships for the Peninsular and Orient Company. He died in Shanghai at the General Hospital there on 29th October 1906, at the same age of 38 as Nellie, and was buried beside her in the Bubbling Well Cemetery. An oval sepia photograph of them standing together survives, both in Edwardian boaters: he nonchalant with a malacca cane, Nellie with a parasol over her shoulder. They stare out at us, almost vibrantly alive, elegant, intelligent, charming, confident, and yet the picture is somehow infinitely sad. It becomes fixed in the memory of all who see it, and in a way encapsulates that vanished species – the sailing ship master and his wife.

Captain and Mrs Sainty.
(Courtesy Edna Leon, Swansea))

I am indebted to Edna Leon of Swansea for the details of George and Susan Sainty.

16.

The Swansea Barque 'Delta'

THE *DELTA* IS A GOOD EXAMPLE of the fine vessels for which Swansea was famous in the latter half of the 19th century. Her eventful career spanned over 33 years before she was sold in the Baltic, and disappears from our sight. The 'Agreement and Account of Crew' and Official Logs covering nearly all of her voyages are lodged in the Swansea City Archives and provided a rich source of material for this chapter.

The *Delta* was built at Cardiff by Hill and Sons and her launch in June 1865 was reported in detail in the *Cambrian* newspaper on 30th of that month as follows:

> "On Wednesday, the upper end of the East Dock and railways adjacent were crowded by thousands of persons anxious to witness the ceremony of launching the first iron sailing vessel built in Cardiff. Scarcely less excitement was felt than was experienced when the first iron steam vessel was launched from the yard of Messrs. Scott Russell into the Taff. The *Delta* is a beautifully shaped iron barque, capable of conveying 1,000 tons of copper ore, for which trade she is intended. Her length is 179 feet, and her extreme breadth 29 feet, depth of hold, 18 feet. She was built by the Messrs. Hill and Sons, of Cardiff, whose yard at the top of the East Dock was thrown open on the occasion of the launch, and was crowded with spectators. The peculiarity in the vessel is that she is not only the first iron sailing vessel to be built in Cardiff, but she is also built entirely of Welsh iron, manufactured for her by the Dowlais Iron Company. When completed she sails from Cardiff with a cargo of coal to the West Coast of America, and returns to Swansea with a cargo of copper ore. She is built for Messrs. Bath and Co., of Swansea. Miss Bath, the niece of the owner, performed the christening ceremony. The launch, which was a side one, was most successfully accomplished amidst the shouts of the spectators and the firing of cannon. The band of the artillery volunteers was present and played 'Rule Britannia' as the vessel plunged into her future element.
>
> To celebrate the event with every possible eclat, Messrs. Hill and

Son invited nearly 300 of the principal merchants of the town and neighbourhood connected in any way with the commercial interest of the town, to a most splendid champagne luncheon, which was exceedingly well served in the modelling room connected with the yard, which was most tastefully decorated for the occasion, in front of which a raised platform had been erected for the band, and another for the ladies and gentlemen invited. After the luncheon had been brought to a close, the chair was taken by Mr. C. S. Hill, who in a very appropriate manner gave the usual loyal and patriotic toasts, which were received with the customary warmth of Welsh people. The 'Bishop and Clergy' was responded to by the Rev. Canon Hawkins of Newport. The rev. gentleman, in the course of his remarks, said he thought it was most appropriate that they should be present on such an occasion, as the bishop and all connected with the established church must feel a deep interest in the success of the commercial interests of the Principality. He hoped the day would soon come when they would see such ships launched into the docks at Newport.

The Rev. Cannon Morgan, of St. Mary's also responded to the same toast, and hoped they were commencing a new era, and that a brighter atmosphere was in store for Cardiff. He hoped it was but the commencement of a trade which would afford a large amount of employment to the Messrs. Hill, Russell, Batchelor, and Hodge, and they would soon rival the other shipbuilding towns on the Clyde, the Thames, and other large rivers.

Mr. Hill, sen., proposed 'The Army, the Navy, and the Volunteers'.

This was responded to by Major Roberts, R.A., Capt. Dornford, and Colonel Clark.

The Chairman then gave – 'Success to the *Delta*'. He thanked the owners for the confidence they had placed in them giving them the entire management of the construction of the vessel, and they intended to complete the undertaking by fitting her up in such a way as to be a credit to her owners. She was not only the first iron sailing vessel built in Cardiff, but she was also the first built of Welsh iron, and every care had been taken in the construction, that she should fulfil all the desires they had respecting her.

Mr. Bath, the ex-mayor of Swansea, briefly responded, and proposed – 'The health of the Builders', which was responded to by the Chairman.

The Chairman gave – 'The health of the Marquis of Bute and his Trustees', which was acknowledged by Mr. Luard.

Mr. Hill then gave – 'Prosperity to the Iron Works of South Wales.'

Colonel Clerk replied in a very humorous address, in which he likened the ironmasters to the farmers, who were always complaining of the badness of the times, though he must confess they were very bad at present for the ironmasters. He also, before sitting down, proposed – 'The Shipping Interest', which was acknowledged by Mr. Batchelor.

'Prosperity to the Trade of Cardiff' – 'The Mayor of Cardiff' – 'Success to the *Delta*' – 'The Captain of the *Delta*' – 'The Foreign Consuls' – 'The Ladies' – and 'The Press', were all subsequently given and responded to, and although the launch took place at one o'clock, the proceedings did not terminate till five o'clock.

The band played a choice selection of pieces during the intervals."

The 'Agreement and Account of Crew' as its name implies, was a legal document authorised by the Board of Trade to define the responsibilities of both employer and crew member. When a crew joined a ship for a voyage they assembled at the Board of Trade Shipping Office, where, under the supervision of the shipping master each man signed against his name and details in the presence of the ship's master. An example of one of these documents is given as Appendix IV and contains such details of the crew member as age, place of birth, last ship, place of engagement, rank on board, wages and place of discharge. It is interesting because the Agreement was opened in Hamburg, as distinct from the majority, which were in Swansea or Liverpool, and illustrates the many changes of crew that could occur during a voyage, and different nationalities that could be found on a Swansea ship.

Associated with the Agreement (commonly known as 'Ships Articles') was the Official Log Book, which was kept during the voyage by the Master, who entered in it any event on board he deemed worthy of recording, usually witnessed by the first mate, as second in command.

The Official Log laid down the 'Nature of the Voyage or Employment', and contained the comprehensive standing instructions laid down by the Board of Trade for the conduct of the ship and all on board, and most importantly, disciplinary measures and regulations among the engagement and discharge of crew members whether in the UK or abroad.

On a vessel's arrival at a foreign port, both these documents had to be deposited with the British Consul or Vice-Consul within 24 hours of arrival (under a penalty of £20 for failing to do so) and would have to be collected immediately before departure.

A good example of the elastic wording used in laying down 'voyage limits' is that for the *Delta*s' voyage 4th September 1875 (Hamburg) to 6th August 1876 at Liverpool:

"Hamburg to the Bristol Channel and from thence to Valparaiso, and/or any other port or places on the West Coast of South America,

Indian Pacific and Atlantic Ocean, China, and Eastern Seas, Australia, Cape-Colonies, Mauritius, West Indies, United States of America, British Provinces, Mediterranean, Black Sea and Baltic, and back to a final port of discharge in the United Kingdom, Continent-of-Europe, calling for orders if required, for a team not to exceed two years."

I have, through the courtesy of the Swansea City Archives Service, obtained a photocopied set of the Ship's Articles and Offical Logs of the *Delta*'s voyages. But since her career was largely spent voyaging to the same places, i.e. between UK/near continent and Chile/Peru, a tabulated list of voyages with dates, places and repetitive details would not be particularly illuminating.

I have instead extracted from the Official Logs entries which speak graphically for themselves and illustrate life on board ship, and what a ship's master had to contend with. As with previous chapters of this work, the choice has been difficult; one could construct a complete book out of the Logs of the *Delta* alone.

The entries in the Official Log were always very clear and precise, and sometimes correct grammar and punctuation is sacrificed to achieve this. I have left Captain Rousey's spelling and punctuation (or rather the lack of it) as I transcribed it. They cast a vivid light on life on board, and above all they illuminate the rigid discipline, harsh conditions and the ever present danger from the sea.

17 JANUARY 1868, AT CHANARAL

John Nash, cook, requested me to give him his discharge, as he was not satisfied with the ship. I immediately granted same and paid him his wages.

George Faulkner Smith, 2nd Mate. Mr. Copus having got killed before he signed the above entry. Mr. Smith the second mate signed in his stead.

28 FEBRUARY 1868, AT VALPARAISO BAY

Frederick Copus, Chief Mate fell down the after hatchway and was killed on the spot. Immediately carried him on deck and sent a boat to H. Bath ship *Clio* which was lying close by for medical assistance when the doctor from that ship came on board he declared Mr. Copus to be dead and passed (sic) all help.

3 MARCH 1868, AT VALPARAISO

Charles Holm seaman has been absent for three days without liberty, and now consider him a deserter.

4 MARCH 1868, AT VALPARAISO

After three days absence Charles Holm has returned on board.

4 MARCH 1868, AT VALPARAISO BAY

List of effects left after Frederick Copus death:

12 pairs pants 3 vests 6 woollen shirts 5 white shirts 3 pairs drawers 3 singlets 6 neck ties
5 pocket handkerchiefs 4 pairs braces 3 belts 5 towels 3 pillow covers 9 hats 4 pair boots
2 Guernsey frocks 3 Duck jackets 2 doz collars Suit of oilskins and sowester 1 brush
2 mufflers 1 bed/quilt 1 blanket 2 sheets 1 pillow 1 canvas bag 15 books 1 sextant
1 case instruments 2 charts 1 pr rulers 1 gold ring 1 watch and chain 1 tapeline 1 album
1 pr opera glasses 1 chest a box containing copper coins and trinkets.

17 MARCH 1868, AT VALPARAISO
Hezekiah Morgan and William Cardew signed the ship's articles the former as Mate the latter as A.B.

10 DECEMBER 1868, LAT 35 48S, LONG 74 50W
John Rutter A.B. lost a case knife overboard ships property for which he will have to pay for £0-1-6.

23 DECEMBER 1868, AT CALDERA
Released John Rutter out of jail he having been on shore since the 20th without liberty. Paid the jailor $7 – seven dollars as a fine. I have to remark that the said John Rutter is a notorious bad character his conduct is such as if he had other parties to join then their conduct would lead to mutiny. On several occasions unknown to me at the time he has insulted the officers most grossly his conduct and refractory manner should be made an example of.

8 FEBRUARY 1868, AT CALDERA
John Rutter incapable and off duty through the effects of liquor. A man employed in his stead for which I paid $1.50.

4 MARCH 1869, AT CALDERA
John Rutter lost a paintbrush overboard which he will have to pay for £0.1.0. He also lost a tin of varnish fall from aloft which was wasted. He will also have to pay for that £0.1.0.

4 JUNE 1869
Read all four entries over to John Rutter in the presence of Mr Walsh Mate. John Rutters reply was that if you try to do your best for me I will do my best for you.

15 OCTOBER 1869, AT VALPARAISO
I came on board asked for the steward found him in a helpless state of intoxication he being what is called dead drunk.

18 OCTOBER 1869, AT VALPARAISO
George Murray drunk and incapable to do duty.

19 OCTOBER 1869, AT VALPARAISO
George Murray drunk and incapable to do duty.

31 OCTOBER 1869, AT VALPARAISO
Gave George Murray liberty to go on shore telling him to be on board at S. Pille as the boat would be on shore at that time.

1 NOVEMBER 1869, AT VALPARAISO
George Murray absent from duty all day.

2 NOVEMBER 1869
-Do-

3 NOVEMBER 1869, AT VALPARAISO
George Murray Steward returned on board. I told him that I had distrated him from the office of steward on board and that in future he would have to do seaman's duty.

16 NOVEMBER 1869, AT VALPARAISO
Came on deck and found the gig missing called all hands found George Murray Steward and Thomas Walkey OS. had left the ship during the night having taken all their effects with them got out another boat went on shore and found the gig in charge of the local authorities.

SAME DAY
Shipped one AB instead of the two above named deserters the ship having been cleared at the consulate before these occurrences he signed the ships articles in the presence of the Chief Mate.

13 FEBRUARY 1870, AT 3.15a.m.
I came on deck and found Charles Stephens Second Mate in charge of the deck asleep on the fore-sky light.

21 APRIL 1870, ABOUT 3 a.m.
Went on deck found Charles Stephens Second Mate in charge of deck asleep. After waiting ten minutes threw some water over him to wake him up when I was about to leave the deck he used some very insolent and threatening language at 8.30 a.m.

called Charles Stephens down in cabin and remonstrated with him on the culpable neglect of going to sleep in charge of the deck also that if caught a third time I would disrate him.

24 APRIL 1870, 3.50 a.m.
Went on deck found Charles Stephens in charge of the deck asleep waited until five minutes past four then rang the bell when he woke up.

18 JULY 1870, AT SOUTHSHIELDS
Z. Olfiskin, Christian Petersen and Hans Lieverts three ABs deserted.

19 JULY 1870, AT SOUTHSHIELDS
Henry Molly cook and steward deserted, having stolen several articles of clothing belonging to me.

26 JULY 1870, AT SOUTHSHIELDS
Charles Johnson OS deserted.

6 AUGUST 1870, OFF LIZARD, S BY 25 MILES
John Jackson not having suited in the capacity of steward since he has been on board put him in the galley to act as cook and took Jacob Logman who acted as cook in the cabin to act as steward dividing their wages equally £3-2-6.

14 AUGUST, 39 23 S, 14 20 W
They people in the forecastle brought their victuals aft and made complaint that it was improperly cooked and not fit to use as human food that since Jackson has been in the galley their victuals had always been spoiled so much that they could not eat it.

16 AUGUST 1870, 35 26 S, 12 32 W
They people brought their victuals aft to let me see that it was improperly cooked and not fit to use. I told them that after another complaint I would disrate the cook that the steward should act in both capacities at the same time chastised the cook in his careless manner of dealing with they peoples.

19 AUGUST 1870, 30 28 S, 13 46 W
This third complaint against John Jackson about the victuals I disrated him to £1-0-0 per month he being entirely useless and not fit to do boys work and also very insolent. Later called Jackson off and read all the former entries over to him his reply was that he never learned to do any work therefore could not do it. I told him he would have to do what he could he refused point blank to do any work also quite saucy.

30 NOVEMBER 1870, AT VALPARAISO BAY

I have to state that John Jackson has shammed sick the greater part of the passage although I am fully convinced there is nothing the matter with him he is being a lazey good for nothing fellow and a great impostor. I can say that he has not done two pounds worth of duty during the time he has been on board the ship the ship having one hand short all the time through his inability to perform the duty he was engaged for. I wish to make this publick so that he will not have the opportunity to impose on any other Ship's Master the same as he has on me he is also very insolent and makes use of very bad language.

REPORT OF THE BRITISH VICE-CONSUL VALPARAISO,
1 DECEMBER 1870

The above named John Jackson was this day brought before me and complained that whilst in the latitude of the Horn, being at the time unable to perform his duty from a bad foot he had been tied to the rigging and left there for nearly two hours exposed to the inclemency of the weather. On my asking Captain Rousey if such was the fact he denied it but my taking up the matter with the crew it appeared that Jackson had actually been punished in the manner stated by order of the mate. I felt it my duty therefore to reprimand Captain Rousey and Mr. George Smith for inflicting a punishment of so serious a nature and 2ndly for not having made an entry specifying what the punishment was in the Log Book. But as I believe the man had been insolent and evidently incompetent I considered it advisable in the interest of all concerned to agree to his discharge. He was accordingly paid off this day by mutual consent and with my sanction.

Alex Gallon
in charge of *Consulate.*

26 MARCH 1871, LAT 48 20 S, LONG 84 30 W

Wind S by E ship heading E to E N E laying too under lower main topsail and fore stay sail. Blowing a fearful gale and high sea running ship labouring heavy and shipping large quantities of water over all a fearful heavy sea swept over the ship filling the decks fore and aft causing the ship to lurch so hevy that all the cargo in the tween decks was thrown over to the Port Side leaving the ship on her beam ends also carrying away all the Top gallant Bullworks from fored to the mizzen rigging also smashing the gig to atoms the long boat and punt with a full set of sucion (sic) and discharging hose belonging to the punt two oars a hold ladder several buckets and scrubbing brushes six capstan bars two watch tackles a gib Boom and Gaff several small spars two fore-castle ladders and various articles about the Decks washed overboard the fore hatch was stove in and all the spars lashed on Deck with waist and stream anchors washed a drift the latter Broken in two pieces the galley was stove in the caboose smashed to atoms and cooking utensils lost the Fore castle hatch having been smashed before it was secured with a sail round it one of the mooring bits was broken in too Everything on deck was washed a drift several of the Ring Bolts were broken and torn out of the Decks the others worked so loose that water was pouring down all over. The Decks were fearfully Ripped up in several places The two spars were so badly mashiated (sic) that they are of little use. About 9 o'clock after the thick of the water and wreck had washed overboard George F. Smith the chief mate along with the crew were trying to secure the spars and fore hatch when the violent Roling of the ship and great quantity of water washing about dashed them about so that the Mate had his left leg broken and right and left ankle badly bruised. He had to be carried aft and put to bed the second mate and two of the crew were also badly bruised leaving only five hands fit to do Duty before the Fore hatch could be got secured a large quantity of water had gone down into the hold. About 10 p.m. all hope of saving the ship being over I directed the helm to be put up and the main yard to be squared in order to get her head before the wind as the only chance of saving her from foundering after a time she paid off at the same time the carpenter was in the hold plugging up the hole where the Ring Bolts had been torn away from. About 2 a.m. after getting the spars secured from doing any more damage Rigged the pumps and pumped her dry out at daylight the wind and sea had abaited a little got the fore hatch off and for the safety of the ship and cargo threw overboard about 30 tons which relieved the ship wonderfully. I have to remark that the running gear was all washed off the pins and so badly out and chaifed that it was rendered quite useless The fore and main hatch Tarpaulins were also torn to shreds and rendered quite useless as soon as duty would alow me to leave the Deck had the laimed people attended too the mates leg set in splints the rest of them their various wounds dressed and bandaged up about 9 a.m. the ship being now out of danger the hands came aft They were then in a very exhausted state after being at work all night. They had wished to know what I intended to do . . .

(Captain Rousey's account carries on for another page and a half, describing how the crew were reluctant to take the vessel round the Horn to Valparaiso. He persuaded them to continue, and eventually brought his ship safely to that port).

ASHORE ON THE GRAVELINES, 4 OCTOBER 1871

On the 30th September 1871 the 'Delta' left Tyne Dock, South Shields, bound for Chile, being towed out by steam tug. Captain Rousey was in command, but as usual on leaving South Shields there was a pilot on board, Hanley W. Chambers. When the tug was cast off a second pilot, John Corbett, of North Shields took charge of the vessel as pilot to the Downs. All went well until Wednesday, 4th October when Captain Rousey made the following entries in his official log:

> 4 OCTOBER 1871
> Strong breeze and hazy weather at 1215 the ship struck on Gravelines sands and thumped hevely (sic) lights E by N which proved to be the

light vesel off Gravelines Sounded the pumps found the ship to be water tight kep the lead going to see if tide rose any more to get ship off if possible About an hour later a fishing boat came alongside and reported that some assistance could be had from Calais Cleared all sails up and stowed them Leaving the ship about 3H 31M the master left to go to Calais to get a steam tug if possible leaving orders to the Mate to get mizzen and after hatches off and throw cargo overboard for next tide When the Master arrived in Calais the steam tug (had) to be telegraphed to London or Dover Tug being employed could not come in at 11am and it was no use wating (sic) any longer Left for the ship in a fishing boat taking extra hands to asist (sic) On board at the same agreeing with the boat to take on anchors and cable out astern for the sum of three hundred franks (sic) which duty they did not perform Continued to throw cargo out of after hold in hopes to get ship off next tide Also got stream anchor and towline ready for steamer At 10 we got a message that steam tug was in Calais Midnight fresh breeze from westward and much rain.

THURSDAY, 5 OCTOBER 1871
Fresh breeze from south west and much rain The steam tug *Palmerston* came to the ship took tow rope on board and continued towing till after high water But to no effect the tide having cut about one and a half feet Afterwards the tug returned to Calais At daylight commenced sending lower royal and top gallant yards and masts and unrigging the ship and the cargo out of after hold Labourers employed from the shore assisting Captain Peare (?) arrived at the ship found the tides cutting fast.

I regret to have to say that in consequence of not having been able to get the assistance of a steam tug on the following tide the ship had to remain on shore hard Had I been able to procure the assistance of a tug I am sure the ship would have been got off in this case.

On arriving at Calais the only one I could find to give me any assistance was a Mr . . . (?) who assisted me to telegraph to Dover Also to engage hands to go to the ship.

The only reason I can give for the ship going on shore when she did was the unsettled state of the atmosphere on Tuesday afternoon there having being (sic) a deal of thunder and lightning during the time the ship was sailing between Lowestoft and Orfordness which must have effected the compasses Also an eastering influence in the water which set the ship considerably to leward of the course then steered as I was on deck all that afternoon and watched the compass carefully and never found the ships head to leward of SW by S and very often as high as SW.

The lights seen at (. . .?) which were taken for Orfordness lights must have been (. . .?) lights which was the cause of the ship going on shore There being a fresh breeze blowing all the first part of the night the ship must have considerably over run her distance Although the pilot was very watchful in every respect in coming along the coast and having taken the ship over once before on a previous voyage I had every confidence in him as a careful trustworthy man.

WEDNESDAY, 11 OCTOBER 1871
After eight days hard labour in throwing cargo overboard and unrigging the ship and other necessary work such as cutting a channel in the sand to start her through succeeded in getting her afloat apparently without damage After getting her off the beach towed her to Calais for the purpose of rigging her and making her seaworthy.

Signed, as all above entries were, by
PETER ROUSEY (Master)
PATRICK KEHOE (Mate)

13 JANUARY 1872, LAT 33 40 S, LONG 41 30 N
Fell in with three boats belonging to the late French barque *L'oriental*, 669 tons, of Captain Gasson of Marseilles from Bordeaux bound to Taita and New Caledonia with general cargo. The three boats contained 38 persons nineteen of which were passengers. They having been 48 howers (sic) in their boats having abandoned their ship two days previous on fire and burning rapidly amongst the passengers four females and two children one two and the other four years old after receiving them all on board the ladies were much exhausted I at once gave up my own room to their accommodation also providing for the gentlemen passengers and ships company as our accommodation would best allow the captain wished me if I choosed to take what I considered most useful of the boats on board for my own use consequently I took one and set the others adrift. Under present circumstances I considered it advisable to steer towards the River Plate in hope to fall in with a vessel bound to some part of the coast which I could transfer them to if not to land them at Montevideo.

(Entry on 17th January records Captain Rousey transferring the survivors to a steamer, *Charrua Viagem*, bound from Montevideo to Rio Gran de Sul).

5 NOVEMBER 1873, AT LIVERPOOL
Delta drawing 17 feet 5 inches water aft and 16 feet 6 inches fored left Liverpool bound for Valparaiso.

8 NOVEMBER 1873
Daniel Olsen who shipped as boatswain and engaged to be on board at 11am after keeping the ship outside in tow of the steamer waiting until 1.30pm shipped William Campbell AB as substitute and proceeded to sea Being very busy not sufficient time to report the same.

9 DECEMBER 1873, LAT 1.46 N, LONG 17 W
Michael Keating os being very incompetent for the duties of an ordinary seaman Considering the high rate of wages he engaged at not being able to steer nor yet find a royal in fact having little of no knowledge whatever I intend to reduce his wages to £1.15.0 instead of £2.5.0 as engaged at.

9 DECEMBER 1873, LAT 46 N, LONG 25 W
Nicholas Heinrich which shipped as Steward being very incompetent and having little or no knowledge of the duties of steward I have had a deal of trouble with him and have threatened to disrate him if he did not improve himself when one month out I found that forty-five days provisions had been consumed through his negligence he is certainly one of the most worthless creatures that I have ever known in the capacity of steward.

11 FEBRUARY 1874, AT VALPARAISO
Discharged Nicholas Heinrich steward by mutual agreement.

12 FEBRUARY 1874, AT VALPARAISO
Shipped William Swan insted (sic) of Nicolas Heinrich discharged as stated above.

13 FEBRUARY 1874, AT VALPARAISO
In the act of heaving up the anchor Christopher Smith AB was keeping a conversation during this time when the Chief Mate told him to cease talking and attend to their work. When the said Christopher Smith said unless he was allowed to talk he would knock off. I called him aft and remonstrated with him when he replied that he wanted to see the Consul if he was not allowed to talk at work.

14 FEBRUARY 1874, AT VALPARAISO
Brought Christopher Smith before the Consul After hearing his statement the Consul ordered him on board and told me the Master if he refuses to go to give him in charge of a vigilante (policeman) I went down to my boat and waited about an hower (sic) and a half seeing nothing of him went on board at 4pm got the ship underweigh and proceeded to Iquique leaving him behind as a deserter.

14 FEBRUARY 1874, AT VALPARAISO
After waiting an hower (sic) and a half for Pierce William Drew Bickerstaffe Apprentice which had left the boat in my absence sent the boat on board at 4am got the ship under weigh and proceeded to Iquique leaving him behind.

26 FEBRUARY 1874, AT IQUIQUE

Thos Mason AB refused to do any more duty Took him ashore before the Consul He had no complaint only that he wanted his discharge The Consul refused his request unless the Master consented to the same and told him to go on board and go to work Failing to do so he would send him to jail Afterwards he made his escape and deserted.

13 MARCH 1874, AT IQUIQUE

Engaged Francis Murphy as Second Mate he having signed the ships articles and commenced duty as such at £7.0.0. per month.

2 APRIL 1874, AT IQUIQUE

Since my arrival at this port I have received advice from Coquimbo from Captain John Roose of the barque *Theta* of Swansea that he liberated Pierce William Drew Bickerstaffe my apprentice out of jail in Valparaiso and that he had him on board of his ship. I wrote Captain John Roose requesting him to send my apprentice down per first passing steamer as I had not treated him as a deserter and that I had all his clothes on board The said apprentice not having arrived and my ship being nearly ready to sail I packed up his effects and sent them on shore in charge of the British Consul to be delivered up to the *Theta* on her arrival at Pisagua.

Since writing the above Pierce William Drew Bickerstaffe has returned to the ship Consequently it is not necessary to comply with the above entries in sending his things on shore in charge of the Consul He has been absent from the ship 46 forty six days.

13 JULY 1874, AT QUEENSTOWN

Pierce William Drew Bickerstaffe having spent four years apprenticeship signs the ships articles today as A.B.

*

Captain Peter Rousey commanded the *Delta* from her first voyage in 1865 until at least 1882, when there is a gap in the Agreements file at Swansea.

In 1886 she was commanded by George F. Smith who had previously sailed as Mate under Captain Rousey. The Agreement which opened 30th November 1887 shows that the ownership had changed to H. J. Madge, Swansea, with Captain Smith still sailing as Master. He was succeeded as Master by William Finch, and thereafter by other Masters until on 19/1/98 the articles closed at Riga, where probably the *Delta* was sold. So far I have not been able to ascertain her final history.

The 1881 Swansea census shows that Captain Rousey lived at 14, Page Street, Swansea, with his wife Charlotte, aged 58, and born in Shetland. There was also a servant, Martha Morgan, aged 63, and a widow. Peter Rousey was born in 1829. His picture appears in a group photograph on page 108.

The barque 'Serena'.
(Swansea Museum)

Swansea pilot cutter 'Tom Rosser', by Henry Birchall, circa 1850.
(Private Collection)

The barque 'Delta'.
(National Museums & Galleries of Wales)

Appendix I

As indicated on page 143 the lists of ships contained in this and Appendix II only include the best known local vessels. A comprehensive list would run into hundreds more and is precluded by space.

Registered tonnages were extracted from the Swansea Shipping Registers or Lloyds Register in the case of those not included in the former. Local tonnages frequently differ from Lloyds because shipowners found they could reduce costs by altering rig and descriptions etc.

LIST OF OWNERS

Bath H. & Son
Beynon W. Llanelly
Bowen John
Burgess J. E.
Burgess, Shaddick & Co.
Goldberg S. & Co.
Hennessy B. R. H.
Hoskin Henry
Jenkins William ('Butcher')
Jenkins David Port Talbot
Madge H. J. & Co.
Meager G. Blaney (includes William)
Mugford & Co.
Nicholson Bros.
Prust J. & Co. (inc. W. H. Prust)
Richards, Power & Co.
Richardson J. C. & Co. (inc. Richardson Bros.)
Rosser Rosser (includes David & John)
Simpson Bros.
Tucker W. H. & Co.
Tulloch Lawrence (includes Gilbert & William)
Swansea Shipping Co. Ltd.
Swansea Merchant Shipowners Co. Ltd.
Vivian & Son

KEY TO ABBREVIATIONS IN LIST

Bk	Barque
Bkn	Barquentine
Bg	Brig
Bn	Brigantine
S	Ship
Scw	Screw (Steamer)
Sr	Schooner
Sw	Snow
Bpl	Barnstaple
Bid	Bideford
Crk	Cork
Dun	Dundee
Fal	Falmouth
Glr	Gloucester
Ham	Hamburg
Hpl	Hartlepool
Hul	Hull
Jer	Jersey
Liv	Liverpool
Lly	Llanelly
Lon	London
Mil	Milford Haven
Mch	Miramichi
Nwc	Newcastle
Npt	Newport
P.E.I.	Prince Edward Island
Sfo	San Francisco
Sws	Swansea
Wtf	Waterford

BATH H. & SON

Name	Rig	Tonnage	Date
Alpha	Bk	398	1864
Beta	Bk	424	1864
Delaware	Bk	339	1859
Delta	Bk	566	1865
Hawkeye	Bk	504	1864
Henry Bath	Bk	489	1863
Iota	Bk	551	1866
Kent	S	445	1862
Mohican	S	357	1859
Pathfinder	Bn	285	1858
Scout	Bk	460	1862
Serena	Bk	493	1866
Swansea	Bk	354	1854
Theta	Bk	509	1865
Tongoy	Bk	484	1866
Uncas	Bk	431	1861
Valentine Helen	Bk	638	1870
Zeta 51109	Bk	558	1865

BEYNON JONES. W. LLANELLY

Name	Rig	Tonnage	Date
Hannah	Bn	199	1875
John Rennie	S	847	1863
Specimen	Bg	220	1856
Reigate	S	1034	?

JOHN BOWEN

Name	Rig	Tonnage	Date
Deerslayer	Bk	347	1854
Guide	Bg	222	1870
Hidalgo	Bn	256	1857
Ida	Bn	199	1873
Jessie	Bk	620	1875
Mary Bowen	Bkn	343	1874
Morna	Bn	230	1873
Ogmore	Bk	328	1865
Ottowa	Bg	232	1874
Raven	Bn	188	1873
Whitehall	Bk	644	1862

BOWRING C. T.

Name	Rig	Tonnage	Date
Hamlet	Bk	1199	1864

BROGDEN ALEXANDER

Name	Rig	Tonnage	Date
Constance Ellen	Bn	171	1871

BURGESS J.

Name	Rig	Tonnage	Date
Admiral Fitzroy	Bk	377	1864
Andromeda	Sr	110	1861
Astracan	Bk	1042	1864
Camena	Bk	585	1865
Clara	Bn	145	1875
Illimani	Bk	590	1875
Lieutenant Maury	Bk	587	1865
Maggie Gross	Bn	185	1869
Patagonia	Bk	299	1863
Pitho	Bn	150	1868
Red Cross Knight	Bk	500	1865
Saint Vincent	S	1377	1872
Standard Bearer	Bk	640	1871
Sedgemoor	Scw Sr	101	?
Southern Belle	Bk	1169	?
Victoria 65543	Sr	110	1861

BURGESS, SHADDICK & CO.

Name	Rig	Tonnage	Date
Abercarne	Bk	735	1875
Aberfoyle	Bk	805	1876
Abernyte	Bk	780	1875
Albany	Bg	270	1871
Araminta	Bk	745	1880
Battle Axe	Bk	629	1874
Cambrian	Bg	192	1872
Colt	B	640	1885
Creole	Bn	192	1870
Gemini	Bk	490	1864
Havre	Sr	177	1864
Highmoor	S	?	1874
Juventa	Bn	272	1855
Maxima	Bk	474	1867
Oliveaga	Scw Sr	412	1872
Oriente	Bk	595	1866
Professor Airey	Bk	448	1866
Thunderbolt	Bk	1193	1885

GOLDBERG S.

Name	Rig	Tonnage	Date
Andaman	Bk	919	1865

Name	Rig	Tonnage	Date
Bertha	Bk	398	1867
Brynymor	S	2131	1902
Candahar	S	1356	1866
Cape Breton	S	1421	1878
City of Sparta	S	1193	1870
City of Madrid	S	1133	1869
Flora	Bn	220	1865
Gateside	Bk	698	1869
Hinda	Bk	443	1869
Isabella	Bg	264	1872
Matilda	Bk	595	1874
Selena	Bg	259	1869
Sydenham	Bk	1062	1870
Vanduara	S	1916	1882

HENNESSY B.R.

Name	Rig	Tonnage	Date
Antarctic	Bk	614	1875
Arctic	Bk	474	1874
Atlantic	Bk	474	1869
Bessie Morris	Bk	486	1860
Georgina	Bk	628	1875
Pacific	Bk	429	1867
Tacna	Bk	403	1857

HOSKIN, HENRY

Name	Rig	Tonnage	Date
Antagonist	Bkn	575	1878
Bridegroom	Sr	497	1874
Bridesmaid	Sr	497	1875
Cwmdonkin	Bn	590	1876
Empress	Bk	405	1873
Harry	Sr	524	1881
H.H.	Bg	235	1868
Juliette	Bk	289	1869
Lizzie	Bg	194	1869
Millie	Bn	242	1866
Swordfish	Sr	139	1885
Symmetry	Bn	199	1866
Titania	Bn	198	1867

JENKINS, DAVID

Name	Rig	Tonnage	Date
David Jenkins	Sr	94	1864

JENKINS, W. ('BUTCHER')

Name	Rig	Tonnage	Date
Catherine Jenkins	Bk	350	?
Diana	Bg	178	1842
Emperor of China	Bk	?	1847
Nepaulese Ambassador	Bk	804	1851
Rajah of Sarawak	S	515	1850

MEAGER, G. BLANEY

Name	Rig	Tonnage	Date
Ariel	Bg	195	1853
Ethel	Bk	372	1863
Kate Helena	Bk	639	1875
Kwasind	Bk	453	1875
Pembroke Castle	Bn	383	1863
Powhattan	Bk	400	1866
Speranza	Bk	446	1869
Stranger	Bk	287	1872
Taunton	Bk	181	1868
W. E. Gladestone	Bk	524	1865
William	Bn	194	1867

MOFFAT, A. G.

Name	Rig	Tonnage	Date
Cheriton	Sr	708	1890
Chislehurst	Bkn	439	1883
Darwin	Bkn	499	1884
Iona	Bkn	319	1882
Reigate	Bkn	398	1884
Vivid	Bkn	423	1885

MOFFAT, ROBERT (GLOUCESTER)

Name	Rig	Tonnage	Date
Curlew	Bkn	280	1874

MUGFORD, G. A.

Name	Rig	Tonnage	Date
Delphi	Bg	224	1866
Lothair	Bn	249	1870
Limena	S	395	1856

NICHOLSON, J.

Name	Rig	Tonnage	Date
Alfred	Bg	253	1871
Delta	Bg	221	1872
Jungfrau	Bk	577	1867

ORD & CO. (SUNDERLAND)

Name	Rig	Tonnage	Date
Hella	Bk	447	1854

PELLIER, W. (LONDON)

Name	Rig	Tonnage	Date
Glance	Bk	887	1869

POWER, S. BROWNING

Name	Rig	Tonnage	Date
Aneroid	Bn	212	1874
Beatrice	Bk	455	1865
Clara	Bn	177	?
Edith May	Bn	196	1865
Emily Anna	Bn	195	1864
Hebe	Bn	214	?
Louisa	Bg	289	1860
Maggie	Bk	324	1867
Maud	Bn	195	1865
Protem	Bg	291	1865

PRUST, J. (AND W. H.)

Name	Rig	Tonnage	Date
Albert T. Young	Bk	306	1874
Annie	Bn	148	1869
Countess	Bg	228	1873
Dilwyn	Bg	271	1874
Ethel	Bn	223	1874
Excel	Bn	209	1871
Gadlys	Bk	500	1874
Kildonian	S	650	1869
Topaz	Bn	196	1870
W. H. Harkness	Bk	481	1870

RICHARDS, POWER & CO.

Name	Rig	Tonnage	Date
Adrian	Bg	240	1871
Artizan	Bn	511	1873
Breeze	Bg	276	1871
Bride	Sr	427	1872
Caspar	Bk	831	1872
Constantine	Bn	245	1872
Emily	Bg	94	1873
Ermina	Bk	631	1877
Evangeline	Bk	532	1874
Fanny	Bg	297	1866
Florence Danvers	Bk	498	1865
Gilpin	Bg	187	1869
Gloster	Bk	303	1871
Hiawatha	Bg	280	1871
Isabelle	Bn	306	1875
Lady Bertha	Scw Sr	460	1877
Lena	Bkn	277	1878
Mabel	Bk	758	1877
Margam	Bg	285	1872
Merlin	Scw Sr	706	1878
One	Bk	523	1870
Resolven	Bg	143	1872
Saint Lawrence	Bk	384	1871
Sir Galahad	Scw Sr	633	1874
Spartan	Bkn	293	1873
Una	Bg	216	1870
Victoria	Bk	748	1874
Warrior	Bk	597	1864
Winifred	Bk	309	1871
WLJ	Bk	290	1870

RICHARDS, T. PICTON

Name	Rig	Tonnage	Date
Eliza	Bg	210	1856
Favourite	Bn	198	?
Florence	Bk	343	1874
Gem	Bg	198	1868
Grace	Bg	270	1866
Hero	Bn	198	1865
Ino	Bg	278	1870
Isabella	Bk	268	1855
Kilvey	Bg	237	1867
Laura	Bg	307	1863
Ninette	Bg	291	1867
Nushka	Bn	417	1875
Pearl	Bn	158	1875
Queen	Bk	435	1865
Richmond	Bg	229	1858
Ripple	Bk	464	1869
Seadog	Bg	248	1870
Sidney A. Jenkins	Bg	259	1868
Smile	Bk	281	1869
Sunshine	Bn	179	1867

RICHARDSON J.C. (AND FAMILY)

Name	Rig	Tonnage	Date
Agnes Blanche	Bk	385	1841
Antonio Vinent	Bk	454	1860
Canadia	Bg	209	1866
Conquerer	S	437	1850
Corsair	Bg	232	1873
Duke of Beaufort	Bk	412	1857
Emily Waters	Bk	404	1861
Faithful	Bn	182	1865
First Lancashire	S	1286	1875
Galatea	Bk	353	1869
Glamorgan	Bk	600	1859
Glanudal	Bk	498	1864
Glendal	Bk	498	1844
Golconda	Bk	499	1866
Hondeklip	Bg	199	1870
Inverness	Bk	724	1869
Lord Marmion	Bk	431	1865
Lynwood	Bn	197	1867
Marquis of Worcester	Bk	419	1861
Marshal Pellissier	S	743	1865
Native Pearl	Bg	192	1865
Ocean Beauty	Bk	578	1868
Ocean King	Bk	419	1859
Ocean Rover	Bk	548	1867
Prince Alfred	S	531	1864
Ravenscraig	S	1267	1866
San José	Bk	393	1861
Sunbeam	Bn	284	1880
Sunbeam	Bg	247	1839
Talca	Bk	420	1866
Victory	Bg	193	1870
Volta	Bk	794	1866
William Davie	Bk	794	1866

ROSSER ROSSER (AND JOHN)

Name	Rig	Tonnage	Date
Avona	Bk	696	1869
Chester	Bg	217	1870
Chevy Chase	Scw Sr	218	1848
Emily	Sr	136	1856
Erycina	Bk	544	1865
Flying Cloud	Paddle Steamer	104	1871
Flying Scud	Paddle Steamer	74	1877
Hollybough	Bk	298	1866
John Rosser	Sr	128	1858
Llewellyn	Bk	498	1875

SIMPSON BROTHERS

Name	Rig	Tonnage	Date
Ben Cruachan	S	1399	1873
Hilda	Sr	100	1873
Janet Cowman	S	1277	1866
Lady Cairns	S	1274	1869
Patterdale	S	1168	1871
Picton Castle	Bk	566	1875
Slieve Race	S	1607	1878
Stately	Bk	373	1849

SWANSEA MERCHANT SHIPOWNERS CO. LTD.

Name	Rig	Tonnage	Date
Epsilon	Scw Sr	126	1879
Eta	Bk	495	1865
Gamma	Bk	556	1865
Helen Burns	S	760	1867
Kappa	Bk	491	1865
Zadok	Bk	579	1868
Zeta 70511	Bk	639	1877

SWANSEA SHIPPING CO. LTD.

Name	Rig	Tonnage	Date
Aldeborough	S	1353	1875
Blonde	Bn	189	1873
Rosewood	Bn	276	1870
Sinbad	Bk	627	1875

TUCKER W. H.

Name	Rig	Tonnage	Date
Alonzo	Bk	317	1872
Blanche	Bg	273	1874
Caradoc	Bn	222	1873
Caswell	Bk	499	1875
Edith	Bg	208	1875
Gladestone	Bkn	297	1873
Hollywood	Bk	286	?
Langland	Bk	639	1875
Osprey	Bk	269	1869
Tawe	Bg	284	1877
W. H.Tucker	Bg	306	1872

TULLOCH L. (AND W.)

Name	*Rig*	*Tonnage*	*Date*
Acacia	Bk	387	1858
Cape Race	Bk	828	1874
Diadem	Bk	285	1873
Fairway	Bk	299	1867
Faith	Bg	261	1865
Fleur de Lys	S	972	1862
Florence	Sw	278	1867
Foxhound	Bk	440	1867
Ianthe	Bk	380	1867
Lord Clyde	Bk	560	1875
Magnet 71836	Sr	92	1875
Magnet 42892	Bn	211	1861
Rose of England	Bk	394	1858
Sketty Belle	Bn	171	1866

TUTTON MORGAN

Name	*Rig*	*Tonnage*	*Date*
Aberlemno	Bk	750	1876
City of Canton	Bk	848	1857
Collingwood	Bg	279	1874
Cosmopolitan	Bk	495	1869
Erica	Bk	443	?
Glen Ariff	Bk	468	1869
Penelope Tutton	Bg	198	1872
Record	Bk	294	1870

VIVIAN H. H.

Name	*Rig*	*Tonnage*	*Date*
Augusta	Sr	125	1849
Fire Opal	Scw Sr	144	1884
Morfa	Sr	299	1862

Appendix II

Name of Ship	*Rig Material*	*Built by*	*Date*	*Tonnage*	*Dimensions Length Breadth Depth*	*Owner*	*Remarks*
Abercarne	Bk Iron	Glasgow McRellar	1875	735	186 31 18.4	Burgess & Shaddick	Captain Lewis mobbed in Swansea for bad treat-ment of crew. *Camb.* 23/8/1878
Aberfoyle	Bk Iron	Southampton Oswald	1876	805	191 32 19	Burgess & Shaddick	Wrecked
Aberlemno	Bk Iron	Dumbarton Birrell Stenhouse & Co.	1876	750	183 31 18	Morgan Tutton	Brought last cargo of copper into Swansea 1905
Abernyte	Bk Iron	Dumbarton Birrell Stenhouse & Co.	1875	780	189 31 18.5	Burgess & Shaddick	
Acacia	Bk	Sunderland	1858	387	125.6 27.6 18	L. Tulloch	Broken up 21/4/1887
Admiral Fitzroy	Bk Wood	Barnstaple Westacott	1864	377	131 27 16	J. E. Burgess	Burnt at sea 3/11/1874
Adrian	Bg Wood	P.E.I. Richards	1871	240	114 25 13	Richards, Power & Co.	Abandoned on voyage Macao to Boston USA 4/4/1881

Name of Ship	*Rig Material*	*Built by*	*Date*	*Tonnage*	*Dimensions Length Breadth Depth*	*Owner*	*Remarks*
Andromeda	Sr Wood	Teignmouth Mansfield	1861	110	84.5 20.7 11	J. R. Burgess	
Agnes Blaikie	Bk Wood	Aberdeen	1841	385	87 20 12	J. C. Richardson	
Aladdin	Bn Wood	P.E.I.	1868	185	104 24 12	H. J. Madge	Lost Huelva Bar 13/2/1870
Albany	Bg Wood	Sunderland	1871	270	116 26.8 16	Burgess & Shaddick	Stranded off Barnstaple 9/12/1887
Albert T. Young	Bk	P.E.I. Young	1874	306	125 27 15	W. H. Prust	
Aldeborough	S Iron	Stockton Richardson & Duck	1875	1353	232 38 23	Swansea Shipping Co. Ltd.	Condemned and Sold By Auction at Montevideo 31/8/1896
Alfred	Bg Wood	P.E.I. Thomas Humphrey	1871	253	111 25 13	J. Nicholson	Wrecked but later re-covered off Bermuda. *Cambrian* 6/5/1881
Alonzo	Bk	P.E.I. McDonald	1872	317	120 26.4 15	W. H. Tucker	
Alpha	Bk Wood	Bridport Cox	1864	398	135 27 17	H. Bath	Lost off Buenos Aires 10/1883
Ambassador	S Later Bk Com-posite	London Walker	1860	714	176 31.5 19	G. Shaddick	Condemned at Port Stanley 27/7/1895
Andaman	Bk Iron	Glasgow John Reid	1865	919	198 33 21	S. Goldberg	

Name of Ship	*Rig Material*	*Built by*	*Date*	*Tonnage*	*Dimensions Length Breadth Depth*	*Owner*	*Remarks*
Aneroid	Bn	P.E.I. Ramsay	1874	212	155 25 13	S. B. Power	
Annie	Bn	P.E.I. Carrol	1869	148	97 24 11	J. Prust	
Antagonist	Bkn Wood	P.E.I. Ramsay	1878	575	170 29 18	Henry Hoskin	
Antarctic	Bk Iron	Glasgow Henderson	1875	614	176 30 17	B. R. Hennessy	
Antonio Vinent	Bk	Sunderland	1860	454	127 20 18	J. C. Richardson	
Araminta	Bk Wood	Nova Scotia	1880	745	158 34 20.5	Burgess & Co.	Wrecked off Port Elizabeth en route Swansea to Algoa Bay 19/1/1889
Arctic	Bk Iron	Glasgow Henderson	1874	616	176 30 17.5	B. R. Hennessy	Foundered 9/1/1879
Ariel	Bg Wood	P.E.I.	1853	195	108 22 12	Richard Meager	
Artizan	Bk Wood	P.E.I. Richards	1873	511	160 29 17	Richards Power & Co.	
Astracan	Bk Iron	Birkenhead Woodside	1864	1042	198 33.5 21.5	Burgess & Co.	Stranded off Melbourne Island South Pacific Oregon to UK
Atlantic	Bk Iron	Glasgow Stephen	1869	474	157.4 27.6 16.8	B. R. Hennessy	

Name of Ship	*Rig Material*	*Built by*	*Date*	*Tonnage*	*Dimensions Length Breadth Depth*	*Owner*	*Remarks*
Augusta	Sr Wood	Swansea Richardsons	1849	125	140 17 10	H. H. Vivian	
Avona	Bk Iron	Stockton on Tees	1869	696	179 30 18	Rosser Rosser	Wrecked Falkland Islands 9/10/1882
Battelaxe	Bk Iron	Glasgow MacFayden	1874	629	181 30	Burgess & Shaddick	Lost at Sea c.6/1881
Beatrice	Bk Wood	Bideford Johnson	1865	455	143 27 17.6	S. B. Power London	
Ben Cruachan	S Iron	Glasgow Barclay Curle	1873	1399	255 37 21	Simpson Bros.	
Bertha	Bk Wood	Sunderland	1867	398	133 28 17	S. Goldberg	
Bessie Morris	Bk Wood	Plymouth	1860	486	154 29 15.5	B. R. Hennessey	
Beta	Bk Wood	Bristol Hill	1864	424	142 28 17	H. Bath	
Blanche	Bg Wood	P.E.I. McDougal	1874	273	121 26 14	W. H. Tucker	
Blonde	Bn Wood	P.E.I. Duncan	1873	189	104 24 12	Swansea Shipping Co. Ltd.	Lost off Bar of Aracata Brazil 25/6/1879
Breeze	Bg Wood	P.E.I. Bollum	1871	276	114 26 14	Richards Power & Co.	Abandoned 30 38E 54 30S
Bride	Sr Wood	P.E.I.	1872	427	148 29 16	Richards Power & Co.	Lost off Cape Horn 1/5/1882

Name of Ship	*Rig Material*	*Built by*	*Date*	*Tonnage*	*Dimensions Length Breadth Depth*	*Owner*	*Remarks*
Bridegroom	Sr Wood (Later Bk)	P.E.I. Ramsay	1874	497	151 29 17	Henry Hoskin	Abandoned at sea on fire 3/9/1884
Bridesmaid	Sr Wood	P.E.I. Richards	1875	497	157 29 17	Henry Hoskin	
Brynymor	S	Dumbarton McMillan	1902	2131		S. Goldberg	See page 127
Caldera	Bk Iron	Sunderland	1861	469	141 27 18	H. J. Madge	Abandoned at sea after collision 8/3/1873
Camena	Bk Iron	Liverpool Vernon	1865	585	171 27 19	J. E. Burgess	
Cambrian	Bg Wood	Swansea Harris	1872	192	100 24 13	Burgess & Shaddick	Lost 1899
Canadia	BG Wood	P.E.I.	1866	209	105 23 12	Richardsons	Wrecked off the coast of Ireland 4/9/1874
Candahar	S Iron	Belfast Harland & Wolf	1866	1356	239 37 23	S. Goldberg	
Cape Breton	S Iron	Clydebank Thompson	1878	1421	239 27 22	S. Goldberg	Wrecked Chile 18/9/1894
Cape Race	Bk Iron	Glasgow	1874	828	197 34 18	L. Tulloch	
Capricorn	Bk Wood	Bideford Cox	1859	389	124 27 17	H. J. Madge	
Caradoc	Bn Wood	P.E.I. Egan	1873	222	116 25 12	W. H. Tucker	

Name of Ship	*Rig Material*	*Built by*	*Date*	*Tonnage*	*Dimensions Length Breadth Depth*	*Owner*	*Remarks*
Caspar	Bk Wood	P.E.I. Richards	1872	831	176 35 20	Richards, Power & Co.	
Caswell	Bk Iron	Dumbarton McMillan	1875	500	156 28 16	W. H. Tucker	Scene of Mutiny see page 159
Catherine Jenkins	Bk Wood		1845	350		W. Jenkins	Lost see page 162
Chelydra	Bk Wood	Newport Mon.	1850	339	152 23 16	H. J. Madge	
Cheriton	Sr Steel	Sunderland Priestman	1890	708	231 32 16	Alexander Moffat	106 HP Steam Screw Schooner
Chester	Bg Wood	P.E.I.	1870	217	104 23 12	Rosser Rosser	Broken up for a hulk at Marseille 19/10/1885
Chevy Chase	Scw Sr Iron	Dumbarton	1848	218	161 24 13	David Rosser	80 HP Steam Screw Schooner
Chislehurst	Bkn Wood	P.E.I. Yeo	1883	439	152 29 16	Alexander Moffat	
City of Canton	Bk Iron	Greenock Steel	1857	848	197 31 20	Morgan Tutton	
City of Sparta	S Iron	Glasgow Stephens	1870	1193	234 35 22	S. Goldberg	
City of Madrid	S Iron	Glasgow Connel	1869	1133	231 35 22	S. Goldberg	
Clara	Bn Wood	P.E.I.		177	94 24 13	S. B. Power	

Name of Ship	*Rig Material*	*Built by*	*Date*	*Tonnage*	*Dimensions Length Breadth Depth*	*Owner*	*Remarks*
Clara	Bn Wood	Novia Scotia	1895	145	90 23.8 21.8	J. Burgess	
Collingwood	Bg Wood	P.E.I. White	1874	279	120 26 14	Morgan Tutton	Abandoned in mid Atlantic 12/11/1877
Colt	Bk Iron	London Langley	1885	640	179 25 18	Burgess & Shaddick	
Conquerer	S Wood	Aberdeen	1850	437	136 26 19	J. C. Richardson	Lost at Long Bay Bahamas 2/1867
Constantine	Bn Wood	P.E.I. Plested	1872	245	116 25 12	Richards, Power & Co.	
Constance Ellen	Bn Wood	Porthcawl	1871	171	103 24 13	Alexander Brogden	
Corsair	Bg Wood	Swansea Richardsons	1873	232	109 24 14	J. C. Richardson	Missing after 9/1878
Cosmopolitan	Bk Iron	Sunderland Doxford	1869	495	151 28 16	Morgan Tutton	
Countess	Bg Wood	P.E.I.	1873	228	117 26 12	John Prust	Wrecked off Fortune Island 3/3/1878
Countess of Bective	Bk Wood	Sunderland	1843	329	?	Nicholson & Son Sunderland	Lost in collision off Lundy Island. En Route Cuba to Swansea *Cambrian* 7/6/1850
Creole	Bn Wood	P.E.I. MacDonald	1870	192	106 25 11	Burgess & Shaddick	

Name of Ship	*Rig Material*	*Built by*	*Date*	*Tonnage*	*Dimensions Length Breadth Depth*	*Owner*	*Remarks*
Cubana	Bk Iron	Sunderland Pile	1862	492	145 27 18	H. J. Madge	
Curlew	Bkn Wood	P.E.I. Richards	1874	280	122 26 14	Robert Moffat Gloucester	
Cwmdonkin	Bn Wood	P.E.I. Richards	1876	590	169 29 17	Henry Hoskin	
Cyprus	Bk Wood	P.E.I. McMullen	1871	349	121 27 14	Henry Hoskin	Wrecked on Towyn Sands 19/5/1871
Darwin	Bkn Wood	P.E.I. Richards	1884	499	161 29 17	A. G. Moffat	
David Jenkins	Sr Wood	Wexford Sparrow	1864	94	87 20 10	David Jenkins	
Deerslayer	Bk Wood	Warrington	1854	347	149 25 19	John Bowen	
Delaware	Bk Wood	Bideford Cox	1859	339	129 26 16	H. Bath	Burnt out at sea off Cape Horn 26/11/1881
Delphi	Bg Wood	P.E.I.	1866	224	110 25 12	G. A. Mugford	Lost 30/3/1874
Delta	Bk Iron	Cardiff Hill	1865	566	162 28 18	H. Bath	See page 212
Delta	Bg Wood	P.E.I.	1872	221	112 24 14	J. Nicholson	Run down by Equateuri off Marseilles 18/2/1879
Diadem	Bk Wood	P.E.I. Ellis	1873	285	116 27 14	L. Tulloch	Abandoned at sea after a collision 12/1/1874

Name of Ship	*Rig Material*	*Built by*	*Date*	*Tonnage*	*Dimensions Length Breadth Depth*	*Owner*	*Remarks*
Diana	Bg Iron	Neath Abbey	1842	178	81 21 13	William Jenkins	Steam Paddle-Wheels
Dilwyn	Bg Wood	P.E.I. McDonald	1874	271	119 25 14	John Prust	
Duke of Beaufort	Bk Wood	Swansea Richardsons	1857	412	118 27 18	Richardsons	Lost off Cuba 2/1865 *Cambrian* 7/4/1865
Edith	Bg Wood	P.E.I. Welsh	1875	208	106 25 14	W. H. Tucker	
Edith May	Bn Wood	P.E.I.	1865	196	93 25 13	S. B. Power	
Eliza	Bg Wood	P.E.I.	1856	196	100 23 12	W. Richards	
Emily	Bg Wood	P.E.I. Bell	1873	94	87 20 10	T. P. Richards	Foundered en route Charleston, S. Carolina to Newcastle 10/1878
Emily 11936	Sr Wood	Bideford Cox	1856	136	91 21 11	John Rosser	
Emily Anna	Bn Wood	P.E.I.	1864	195	94 24 13	S. B. Power	Sailed from Cadiz for Swansea 18/1/1870 not heard of since
Emily Waters	Bk Iron	Liverpool Miller	1861	446	153 25 17.5	Richardsons	
Emperor of China	Bk Wood	Bideford Cox	1847	485		William Jenkins	

Name of Ship	*Rig Material*	*Built by*	*Date*	*Tonnage*	*Dimensions Length Breadth Depth*	*Owner*	*Remarks*
Epsilon	Sr Steel	Newcastle Mitchell	1879	126	126 29 11	Swansea Merchant Shipowners Co. Ltd.	Twin Scw Steam 70 H.P.
Ermina	Bk Wood	P.E.I. Yeo	1877	631	158 32 17	Richards, Power & Co.	
Erica	Bk Wood	P.E.I. Plested		443	140 29 17	Morgan Tutton	
Erycina	Bk Wood	Southwick Nicholson	1865	544	152 29 20	Rosser Rosser	Foundered en route S. Carolina to Birkenhead 12/1878
Eta	Bk Iron	Bristol Hill	1865	495	151 29 18	Swansea Merchant Shipowners Co. Ltd.	
Ethel	Bk Iron	Sunderland Rawson	1863	372	119 25 17.5	G. B. Meager	
Ethel	Bn Wood	P.E.I. Coffin	1874	223	116 25 12	John Prust	
Evangeline	Bk Wood	P.E.I. Richards	1874	532	146 32 18	Richards, Power & Co.	
Empress	Bk Wood	P.E.I. Keefe	1873	405	132 27.7 16.9	Henry Hoskin	
Excel	Bn Wood	P.E.I.	1871	209	104 24 12	John Prust	Abandoned 14/4/1874
Faith	Bg Wood	Quebec	1865	261	111 25 12	L. Tulloch	Lost Casampa Bay 24/8/1873

Name of Ship	*Rig Material*	*Built by*	*Date*	*Tonnage*	*Dimensions Length Breadth Depth*	*Owner*	*Remarks*
Fairway	Bk Wood	Sunderland	1867	299	114 26 16	L. Tulloch	Abandoned at sea 9/8/1887
Faithful	Bn Wood	P.E.I.	1865	182	100 24 12	Richardson	
Fanny	Bg Wood	P.E.I. Ellis	1866	297	113 26.4 15.4	Richards, Power & Co.	
Favourite	Bn Wood	P.E.I.		198	104 23 12	T. P. Richards	
Fire Opal	Sr Iron	Cubitt Town	1884	144	152 22 12	H. H. Vivian	Scw Sr Steam 68 H.P.
First Lancashire	S Iron	Durham Osborne & Graham	1875	1286	222 36 22	J. C. Richardson	
Fleur de Lys	S	Whiteinch Lanarkshire	1862	972	199 33 21	L. Tulloch	Converted to a hulk at Brisbane 23/9/1907
Flora	Bn Wood	P.E.I	1865	220	109 20 13	S. Goldberg	Lost off Lowestoft 31/1/1870
Florence Danvers	Bk Wood	Bideford Johnson	1865	498	141 28 18	Richards, Power & Co.	
Florence 71425	Bk Wood	P.E.I. Plested	1874	343	127 28 16	T. P. Richards	
Florence 58084	Sw	Sunderland Gill	1867	278	110.5 27 15.8	W. Tulloch	
Foxhound	Bk Iron	Hartlepool Denton Gray	1867	440	147 27 17	L. Tulloch	

Name of Ship	*Rig Material*	*Built by*	*Date*	*Tonnage*	*Dimensions Length Breadth Depth*	*Owner*	*Remarks*
Flying Cloud	Paddle Steamer Iron	Newcastle Hepple	1871	104	96.5 18.1 9.2	Rosser Rosser	
Flying Scud	Paddle Steamer	Neath Jones	1877	74	83 18.1 8.6	Rosser Rosser	
Gadlys	Bk Iron	Dumbarton Birrell	1874	500	160 27 17	J. Prust	
Galatea	Bk Wood	P.E.I. Bourke	1869	353	138 27.6 15	Richardson	
Gamma	Bk Wood	Bideford Cox	1865	556	153 29 18.7	Swansea Merchant Shipowners Co. Ltd.	
Gateside	Bk Iron	Dumbarton MacMillan	1869	698	184 29 18.5	S. Goldberg	
Gem	Bg Wood	P.E.I.	1868	198	109 24 12	T. P. Richards	Lost off Newfound land Banks 5/6/1870
Gemini	Bk Wood	Jersey	1864	490	187 27 17	Burgess, Shaddick & Co. Ltd.	
Georgina	Bk Iron	Hartlepool Gray	1875	628	186 29.8 17.6	B. R. Hennessy	
Gilpin	Bg Wood	Swansea Richards	1869	187	100 30 12.7	Richard, Power & Co. Ltd.	
Glamorgan	Bk Wood	Bideford Cox	1859	600	151 29.6 190	Richardson	
Glance	Bk Iron	Sunderland Watson	1869	887	197.7 33 20.8	W. Pellier London	

Name of Ship	*Rig Material*	*Built by*	*Date*	*Tonnage*	*Dimensions Length Breadth Depth*	*Owner*	*Remarks*
Glenariff	Bk Iron	Dumbarton MacMillan	1869	468	161 28 17	Morgan Tutton	
Glendal	Bk Wood	Sunderland	1844	498	141 29 18	J. C. Richardson	
Glenrafon	BK Iron	Dundee Brown	1866	472	156 26.6 17	Richardson	
Glanudal	Bk Iron	Sunderland Robinson	1864	498	141 29.6 18.4	Richardson	
Gladestone	Bkn Wood	P.E.I. Hyndman	1873	297	173.6 25.6 14.7	W. H. Tucker	
Gloria	Bk Iron	Sunderland Doxford	1868	735	172 30 19.3		
Gloster	Bk Wood	P.E.I. Bollum	1873	303	122 27 14	Richards, Power & Co.	
Golconda	Bk Iron	Dundee Brown	1866	499	159 26.7 17.3	Richardson	
Grace	Bg Wood	P.E.I.	1866	270	107 25 15	T. P. Richards	
Guayacan	Bk Wood	Sunderland	1857	451		H. J. Madge	Wrecked
Guide	Bg Wood	P.E.I.	1870	222	106 24 12	John Bowen	Wrecked near Gothenburg 14/10/1875
Hamlet	Bk Iron	London Rennie	1864	1199	215 34 22	C. T. Bowring	
Hannah	Bn Wood	P.E.I. Duncan	1875	199	103 24.5 12.9	William Beynon Llanelly	

Name of Ship	*Rig Material*	*Built by*	*Date*	*Tonnage*	*Dimensions Length Breadth Depth*	*Owner*	*Remarks*
Harry	Sr Wood	P.E.I. Richards	1881	524	165 28 17	Henry Hoskins Jnr.	
Havre	Sr Iron	Glasgow	1864	177	147 21 12	Burgess & Shaddick	
Hawkeye	Bk Wood	Bideford Cox	1864	504	142 29 18	H. Bath	Later became Bn Alcibades
Hebe	Bn Wood	P.E.I.		214	99 26 12	S. B. Power	
Hecla	Bk Wood	Hartlepool	1854	447	136 24 17	Ord & Co. Sunderland	Brought Yellow Fever to Swansea *Cambrian* 17/3/1871 24/3/1871
Helen Burns	S Iron	Glasgow Duncan	1867	760	193 32 19	Swansea Merchant Shipping Co. Ltd.	
Henry Bath	Bk Wood	Bideford Cox	1863	489	141 29 18	H. Bath	
Hero	Bn Wood	P.E.I.	1865	198	94 20 13	T. P. Richards	Wrecked in River Seine
Herradura	Bk Iron	Sunderland	1862	499	148 27 17	H. J. Madge	Sailed for Cape Horn 1880. Not heard of since
Hiawatha	Bg Wood	P.E.I. Bollum	1871	280	116 26 14	Richards, Power & Co. Ltd.	
Hidalgo	Bn Wood	Bristol	1857	256	120 24 15	John Bowen	

Name of Ship	*Rig Material*	*Built by*	*Date*	*Tonnage*	*Dimensions Length Breadth Depth*	*Owner*	*Remarks*
Highmoor	S (later Bk) Iron	Sunderland Thompson	1874		207 35 22	Burgess & Shaddick	Stranded Glorice Reef Madagascar 19/1/1895
Hilda	Sr Wood	Barnstaple Westacott	1873	100	87 22.6 11	Simpson Bros.	
Hinda	Bk Iron	Llanelly Nevill	1869	443	157 27 16	S. Goldberg	
H. H.	Bg Wood	P.E.I.	1868	235	108 25 14	Henry Hoskin	Condemned as unsea-worthy 2/3/1877
Hondeklip	Bg Wood	Swansea Richardsons	1870	199	106 24 14	J. C. Richardson	Lost on Goodwin Sands 18/2/1894
Hollybough	Bk Iron	Durham Briggs	1866	298	114 27 16	J. Rosser	
Hollywood	Bk Wood	P.E.I. Waite		286	121 26 14	W. H. Tucker	Abandoned 37W 58S 24/5/1877
Huasco	Bk Wood	Bideford Cox	1860	435	132 27 17	H. J. Madge	
Ianthe	Bk Iron	Dundee	1867	380	135 26 16.6	L. Tulloch	
Ida	Bn Wood	P.E.I. McLaughlin	1873	199	105.4 24 12.9	W. Bowen Jnr.	
Illimani	Bk Iron	Glasgow Dobie & Sons	1875	590	174 29 17	J. E. Burgess	
Ino	Bg Wood	P.E.I. Bollum	1870	278	115 26 14	T. P. s Richards	Abandoned 14/12/1876

Name of Ship	*Rig Material*	*Built by*	*Date*	*Tonnage*	*Dimensions Length Breadth Depth*	*Owner*	*Remarks*
Inverness	Bk Iron	Aberdeen Hall, Russell & Co.	1869	724	177 32 18	J. C. Richardson	Stranded and lost near Para 20/12/1896
Iona	Bkn Wood	P.E.I. Yeo	1882	319	121 27 14	A. G. Moffat	
Iota	Bk Wood	Bideford Cox	1866	551	151 29 19	H. J. Bath	
Isabella	Bg Wood	P.E.I. Richards	1872	264	114 27 14	S. Goldberg	Condemned and sold in Savannah 1/11/1887
Isabella	Bk Wood	P.E.I.	1855	268	111 24 15	T. P. Richards	
Isabelle	Bn Wood	P.E.I. Richards	1875	306	122 27 14	Richards, Power & Co.	Lost 1895
Janet Cowman	S (later Bk) Iron	Greenock Steele & Co.	1866	1277	224 35 22	Simpson Bros.	Abandoned 57 51 S 66 37 W 17/4/1890
Jessie	Bk Iron	Hartlepool Gray	1875	620	176 29 18	John Bowen	Lost off Cape Horn 19/5/1879
John Rennie	S Iron	Greenwich	1863	847	177 32 20	W. Beynon Jones	
John Rosser	Sr Wood	Barnstaple Westacott	1858	128		J. Rosser	
Julie	Bk Iron	Sunderland Ilif & Mounsay	1870	490	168 26 16	G. Shaddick	Wrecked near Rio Grande de Sul 6/6/1889 (Newport to Rosario)

Name of Ship	*Rig Material*	*Built by*	*Date*	*Tonnage*	*Dimensions Length Breadth Depth*	*Owner*	*Remarks*
Juliette	Bk Wood	P.E.I. Gillis	1869	289	120 26 14	Henry Hoskin	
Jungfrau	Bk Iron	Sunderland Doxford	1867	577	158 30 18	J. Nicholson	
Juventa	Bn Wood	Holland	1855	272	102 25 15	Burgess & Shaddick	Abandoned off Morocco 5/12/1874
Kappa	Bk Iron	Glasgow Stephen	1865	491	156 27 17	Swansea Merchant Shipowners Co. Ltd.	Swansea to Valparaiso in 71 days
Kate Helena	Bk Iron	Glasgow Stephen	1875	639	170 29 18	G. B. Meager	
Kent	S Wood	Sunderland	1862	445	139 28 17	H. Bath	
Kildonian	S Iron	Glasgow Stephen	1869	650	177 30 18	J. Prust	
Kilvey	Bg Wood	P.E.I.	1867	237	105 24 15	T. P. Richards	
Kwasind	BK Wood	P.E.I. Bell	1875	453	145 29 17	G. B. Meager	
Lady Bertha	Scw Sr Iron	Newcastle Schlesinger	1877	460	186 26.7 15.7	Richards, Power & Co.	
Lady Cairns	S Iron	Belfast Harland & Wolf	1869	1274	216 35 22	Simpson Bros.	
Langland	Bk Iron	Middles-borough Dixon	1875	639	173.6 29.6 17.6	W. H. Tucker	See page 97

Name of Ship	*Rig Material*	*Built by*	*Date*	*Tonnage*	*Dimensions Length Breadth Depth*	*Owner*	*Remarks*
Laura	Bg Wood	P.E.I.	1863	307	113 26 15	T. P. Richards	Foundered off Portland 16/1/1870
Lena	Bkn Wood	P.E.I. Carroll	1878	277	123.4 25.6 14.9	Richards, Power & Co.	
Lieutenant Maury	Bk Wood	Barnstaple Westacott	1865	587	135 27 17	J. Burgess	Burnt out while loading at Port Nolloth
Limena	S Wood	Hartlepool	1856	395	132 27 17	Alfred Mugford	
Lizzie	Bg Wood	P.E.I. Herriot	1869	194	108 24 13	Henry Hoskin	Run down off Scilly 25/2/1872
Llewellyn	Bk Iron	Glasgow Stephen	1875	498	162 27.6 17.1	John Rosser	
Lord Clyde	Bk Iron	Glasgow Stephens	1875	560	173.4 28 17.6	L. Tulloch	
Lord Marmion	Bk Wood	Swansea Richardson	1865	431	133.4 27.6 18.7	Richardson	
Lothair	Bn Wood	P.E.I. McDonald	1870	249	120 24.8 12.9	Mugford & Co.	
Louisa	Bg Wood	P.E.I.	1866	289	114 26 14	S. B. Power	Abandoned 45 N 22.45 W 20/12/1871
Lynwood	Bn Wood	P.E.I. Hazard	1867	197	104.6 24 12.9	Richardson	
Mabel	Bk Wood	P.E.I. Bollum	1877	758	178.5 34 18.6	Richards, Power & Co.	

Name of Ship	*Rig Material*	*Built by*	*Date*	*Tonnage*	*Dimensions Length Breadth Depth*	*Owner*	*Remarks*
Maggie	Bk Wood	P.E.I. Ellis	1867	324	123.5 26.5 15.3	S. B. Power	
Maggie Gross	Bn Wood	New Brunswick Hine Bros.	1869	185	101.8 25.9 11.4	J. E. Burgess	Stranded and became a total loss near Port Charlotte 14/12/1880
Magnet 42892	Bn Wood	P.E.I.	1861	211	100.5 23.6 12.5	L. Tulloch	Swansea to Valparaiso in 69 days
Magnet 71836	Sr Wood	Sydney MacGowrie	1875	92	87 20.7 8.6	L. Tulloch	
Margam	Bg Wood	P.E.I.	1872	285	120 29 14	Richards, Power & Co.	
Marquis of Worcester	Bk Wood	Swansea Richardson	1861	419	124 27 18	J. C. Richardson	
Marshall Pellissier	S Wood	Quebec	1865	743	164 32 20	J. C. Richardson	
Martha	Bn Wood	Shoreham	1867	198	108 26 12	Vivian Stevens St. Ives Cornwall	Wrecked 9 miles off Calais 20/12/1877
Mary Bowen	Bkn Wood	P.E.I. Matheson	1874	343	132 29 14	John Bowen	
Matilda	Bk Wood	P.E.I. Richard	1874	595	154.8 32 17.7	S. Goldberg	
Maud	Bn Wood	P.E.I.	1865	195	94 23 13	S. B. Power	Foundered off Portland 4/12/1872
Maxima	Bk Iron	Newcastle Palmers	1867	474	153.6 28.4 17.4	Burgess & Shaddick	

Name of Ship	*Rig Material*	*Built by*	*Date*	*Tonnage*	*Dimensions Length Breadth Depth*	*Owner*	*Remarks*
Merlin	Scw Sr Iron	Wallsend Schlesinger	1878	70	225 30 18	Richards, Power & Co. Ltd.	120 H.P. Steam Screw Sr
Millie	Bn Wood	P.E.I. Stewart	1866	242	117 24 12	Henry Hoskins	
Mohican	S Wood	Waterford	1859	357	130 26 16	H. Bath	
Morna	Bn Wood	P.E.I. Duncan	1873	230	117 26 12	John Bowen	
Morfa	Sr Iron	Newcastle Palmer	1862	299	140 25 14	Vivian & Son	Foundered off Strumble Head 19/2/1894
Native Pearl	Bg Wood	P.E.I.	1865	192	106 23 12	J. Richardson	
Nepaulese Ambassador	Bk Wood	Bideford Cox	1851	804		W. Jenkins	
Ninette	Bg Wood	P.E.I.	1867	291	112 26 15	T. P. Richards	
Nushka	Bn Wood	P.E.I. Carroll	1875	417	138 29 17	T. P. Richards	
Ocean Beauty	Bk Iron	Durham	1868	578	154 29 18	J. Richardson	Stranded on Aberavon sands and wrecked 2/12/1886
Ocean King	Bk Wood	Sunderland Laing	1859	419	131 26 16	J. Richardson	Foundered 29 36S 16 52E 25/1/1881
Ocean Rover	Bk Com- posite	Durham	1867	548	153 28 18	J. Richardson	Destroyed by fire at Laraquele 20/8/1891

Name of Ship	*Rig Material*	*Built by*	*Date*	*Tonnage*	*Dimensions Length Breadth Depth*	*Owner*	*Remarks*
Ogmore	Bk Wood	Sunderland Thompson	1865	328	117 27 17	John Bowen	
Oliveaga	Scw Sr Iron	Sunderland Bertram & Haswell	1872	412	194 26 13	Burgess & Shaddick	Lost off Sables de Mohne 15/11/1884
One	Bk Wood	P.E.I. Richards	1870	523	143 30 18	Richards, Power & Co.	Broken up at Cardiff 24/4/1888
Oriente	Bk Iron	Liverpool Vernon	1866	595	171.6 27.9 18	Burgess & Co.	Wrecked 1893
Osprey	Bk Wood	P.E.I. Hyndman	1869	269	119 25 14	W. H. Tucker	Sold in Bombay 19/5/1879
Ottawa	Bg Wood	P.E.I. Gillow	1874	232	116 26 12	John Bowen	Lost off Para, Brazil 31/10/1879
Pacific	Bk Iron	Glasgow	1867	429	146 26 16	B. R. Hennessy	Lost 1893
Patagonia	Bk Wood	Barnstaple Westacott	1863	299	122 25 15	J. Burgess	Wrecked at Madeira 6/1876
Pathfinder	S (later Bn) Wood	Waterford	1858	285	131 26 14	H. Bath	
Patterdale	S Iron	Whitehaven	1871	1168	224 36 21	Bernard Simpson	Sailed from Newcastle for Iquique 24/3/1902, not heard of since
Pembroke Castle	Bn Iron	Glasgow	1863	383	147 26 16	Meager	
Pearl	Bn Wood	P.E.I. Richards	1875	158	98 23 12	T. P. Richards	

Name of Ship	*Rig Material*	*Built by*	*Date*	*Tonnage*	*Dimensions Length Breadth Depth*	*Owner*	*Remarks*
Penelope Tutton	Bg Wood	P.E.I. Miller	1872	198	111 24 13	Morgan Tutton	Abandoned at sea 45 39N 28 30W 20/1/1881
Picton Castle	Bk Iron	Glasgow Stephen	1875	566	173 28 17	Simpson Bros.	Wrecked 1895
Pitho	Bn Wood	Swansea Meager	1868	150	96 22 12	J. Burgess	
Powhattan	Bk (later Bk)	Swansea Meager	1866	400	130 27 17	G. B. Meager	
Prince Alfred	S Iron	Queenstown Wheeler	1864	531	142 30 12	J. C. Richardson	
Professor Airey	Bk Wood	Barnstaple Westacott	1866	448	139 28 17	J. Burgess	
Protem	Bg Wood	P.E.I.	1865	291	115 26 14	S. B. Power	Lost in ice St. Lawrence River 31/12/1887
Queen	Bk Wood	P.E.I.	1865	435	159 26 17	T. P. Richards	Lost at Tilt Cove
Rajah of Sarawak	S	Bideford Cox	1850	515		William Jenkins	See page 62
Raven	Bn Wood	P.E.I. Duncan	1873	188	106 24 12	John Bowen	
Ravenscraig	S Iron	Greenock	1866	1267	219 35 22	J. C. Richardson	
Record	Bk Wood	P.E.I.	1870	294	119 28 14	Morgan Tutton	Damaged at Ceuta, sold by auction 17/10/1878

Name of Ship	*Rig Material*	*Built by*	*Date*	*Tonnage*	*Dimensions Length Breadth Depth*	*Owner*	*Remarks*
Red Cross Knight	Bk Wood	Barnstaple Westacott	1868	500	134 20.5 17	J. E. Burgess	Went ashore while towing up to Cardiff. Total loss
Reigate	Bkn Wood	P.E.I. Yeo	1884	398	145 27 15	A. G. Moffat	
Reigate	S Iron	Deptford Kingsley		1034	210 32 21	W. Beynon Jones	
Resolven	Bg Wood	P.E.I. Richards	1872	143	96 20 11	Richards, Power & Co.	
Richmond	Bg Wood	P.E.I.	1858	229	104 24 15	T. P. Richards	
Ripple	Bk Wood	P.E.I.	1869	464	139 29 17	T. P. Richards	Abandoned off Cape Breton 27/11/1887
Rose of England	Bk Wood	Hartlepool	1858	394	193 27 17	L. Tulloch	
Rosewood	Bn Wood	P.E.I. White	1870	276	117 26 12	Swansea Merchant Shipping Co. Ltd.	Missing since 27/7/1875
Saint Lawrence	Bk Wood	P.E.I. Richards	1871	384	128.6 29 16	Richards, Power & Co.	
Saint Vincent	S Wood	Quebec Dunn	1872	1377	200 36 24.8	J. E. Burgess	
San José	Bk Wood	Sunderland Pickersgill	1861	393	131 28 18	J. C. Richardson	Condemned at Lagos and sold 10/1888
Scout	Bk Wood	Bideford Cox	1862	460	137 28 18	H. Bath	

Name of Ship	*Rig Material*	*Built by*	*Date*	*Tonnage*	*Dimensions Length Breadth Depth*	*Owner*	*Remarks*
Seadog	Bg Wood	P.E.I. Ramsay	1870	248	108 25 14	T. P. Richards	Foundered 47 52 N 57 50 W 10/5/1871
Sedgemoor	Scw Sr Iron	Hartlepool Richardson		101	106 20 8.4	Burgess & Shaddick	15 HP Scw Sr Steam
Selena	Bg Wood	P.E.I. Coffin	1869	259	113 25 14	S. Goldberg	Foundered 15 miles off Jamaica 12/3/1880
Serena	Bk Com- posite	Sunderland	1866	493	154 27 18	H. Bath	Lost 58 34 S 69 50 W 3/8/1893 See page 144
Sidney A. Jenkins	Bg Wood	P.E.I. MacDonald	1868	259	111 26 14	T. P. Richards	Condemned & sold at St. Thomas 15/12/1882
Sinbad	Bk Iron	Sunderland Thompson	1875	627	186 29 17	Swansea Shipping Co. Ltd.	Last vessel to load for Chile. Sailed 1/4/1895 *Sea Breezes* May 1939 p42
Sir Galahad	Scw Sr Iron	Whitby Turnbull	1874	633	224 30 16	Richards, Power & Co.	
Sketty Belle	Bn Wood	Quebec	1866	171	102.8 22.6 12.7	L. Tulloch	
Slieve Race	S Iron	Belfast Harland & Wolf	1878	1607	257 68 23	Simpson Bros.	
Smile	Bk Wood	P.E.I. Richards	1869	281	116 26 14	T. P. Richards	Abandoned 44 N 42 W 12/9/1875
Southern Belle	Bk Iron	Bristol Hill		1169		Burgess & Shaddick	Abandoned 39 30 N 09 3 W 15/11/1891

Name of Ship	*Rig Material*	*Built by*	*Date*	*Tonnage*	*Dimensions Length Breadth Depth*	*Owner*	*Remarks*
Spartan	Bkn	P.E.I. Richards	1873	293	120 27 14	Richards, Power & Co.	Abandoned 40 54 N 32 16 W 18/1/1882
Specimen	Bg Wood	P.E.I.	1858	220	108 24 12	John Beynon	Lost off Caldy 3/1865
Speranza	Bk Wood	Sunderland Nicholson	1869	446	138 29 17	G. B. Meager	Abandoned on passage Betts Cove to Swansea 23/11/1877
Standard Bearer	Bk Wood	Barnstaple Westacott	1871	640	145 20.6 18.5	J. E. Burgess	Abandoned off Brazilian coast 5/6/1881
Stately	Bk Wood	Hartlepool Grey	1849	373	108 25 18	Simpson Bros.	
Stranger	Bk Wood	Swansea Meager	1872	287	121 26 15	G. B. Meager	Burnt out 45 miles off Port Nolloth 27/8/1878
Sunbeam	Bn Wood	P.E.I. Yeo	1880	284	148 28 15	J. C. Richardson	Lost off Puerto Caballo 28/3/1884
Sunbeam	Bg Wood	Stockholm	1839	247	88 29 16	J. C. Richardson	
Sunshine	Bn Wood	P.E.I.	1867	179	97 24 12	T. P. Richards	Abandoned 23/2/1877
Swansea	Bk Wood	Sunderland	1854	354	121 27 7	H. Bath	
Swordfish	Sr	P.E.I Richards	1885	139	114 21 10	Henry Hoskin	Stranded off Trinidad 21/8/1881

Name of Ship	*Rig Material*	*Built by*	*Date*	*Tonnage*	*Dimensions Length Breadth Depth*	*Owner*	*Remarks*
Sydenham	Bk Iron	Glasgow Stephen	1870	1062	210 34 21	S. Goldberg	
Symmetry	Bn Wood	P.E.I.	1866	199	106 24 12	Henry Hoskins	Lost on passage from Combrero 22/9/1876
Tacna	Bk Wood	Liverpool	1857	403	139 26 18	B. R. Hennessy	Burnt at sea 11 N 26 W 11/9/1885
Talca	Bk Iron	Cardiff Batchelor	1866	420	142 26 6	J. Richardson	
Taunton	Bk Com-posite	Hartlepool Denton Grey	1868	181	181 32 17	G. B. Meager	
Tawe	Bg Wood	P.E.I. McDougal	1877	284	123 26 14	W. H. Tucker	
Theta	Bk Wood	Bideford Cox	1865	509	149 29 18	H. Bath	
Thunderbolt	S (Later Barque) Iron		1885	1193	210 35 23	Burgess & Shaddick	Lost on Martwell Reef, Bonavista Cape Verde 3/11/1893
Titania	Bn Wood	P.E.I.	1867	198	105 24 12	Henry Hoskin	Lost 50 miles off Cape Clear
Tongoy	Bk Wood	Sunderland	1866	484	153 27 18	H. Bath	Lost off Drogheda 3/3/1870
Topaz	Bn Wood	P.E.I.	1870	196	105 24 12	John Prust	
Una	Bg Wood	P.E.I. Richards	1870		114 26 12	Richards, Power & Co.	Lost on Scatter Island, Cape Breton 20/6/1871

Name of Ship	*Rig Material*	*Built by*	*Date*	*Tonnage*	*Dimensions Length Breadth Depth*	*Owner*	*Remarks*
Uncas	Bk Wood	Bideford Cox	1861	431	134.8 28.4 17.6	H. Bath	
Valentine Helen	Bk Iron	Glasgow	1870	638	176 29 17	H. Bath	
Vanduara	S Iron	Glasgow Connell	1882	1916	291 42 24	S. Goldberg	Sunk in World War I
Victoria	Bk Wood	P.E.I. Richards	1874	748		Richards, Power & Co.	
Victoria 65543	Bk Iron	Sunderland	1867	679	167 29 19	J. E. Burgess	
Victory 59798	Bg Wood	P.E.I. Miller	1870	193	99 24 12	J. Richardson	
Volta	Bn Wood	P.E.I. MacDonald	1870	221	105 24 12	J. Richardson	Abandoned at sea 15/3/1874
Vivid	Bkn	P.E.I. Yeo	1885	423	145 30 14	A. G. Moffat	
Warrior	Bk Wood	P.E.I Balls	1864	597	140.6 32.2 19.7	Richards, Power & Co.	
W. E. Gladestone	Bk Com-posite	Sunderland Peverill	1865	524	148 30 17	G. B. Meager	Abandoned 800 miles off Cape Horn 5/9/1881
W. H. Harkness	Bk Com-posite	Middles-borough	1870	481	145 28 18	W. H. Prust	
W. H. Tucker	Bg Wood	P.E.I.	1872	306	120 26 14	W. H. Tucker	

Name of Ship	*Rig Material*	*Built by*	*Date*	*Tonnage*	*Dimensions Length Breadth Depth*	*Owner*	*Remarks*
Whitehall	Bk Wood	Devon	1862	844	183 32 22	John Bowen	Wrecked near Willemstad, Holland 7/2/1883
William	Bn Wood	P.E.I.	1867	194	104 20 12	G. B. Meager	Lost in Belfast Lough 1/1/1875
William Davie	Bk Com-posite	Helensburgh	1866	794	187 31 19	J. C. Richardson	Burnt out at sea 10 miles off Diego Ramares 17/10/1889
Winifred	Bk Wood	P.E.I. Ramsay	1871	309	119 26 14	Richards, Power & Co.	
WLJ	Bk Wood	P.E.I. Ramsay	1870	290	120 26 14	Richards, Power & Co.	
Zadoc	Bk Iron	Seacombe	1868	579	173 28 18	Swansea Merchant Shipowners Co. Ltd.	Sailed Ensenada 1/6/1868 not heard of since
Zeta 51109	Scw Bk Com-posite	Glasgow	1865	558	191 28 17	H. Bath	90 HP Steam Scw. Sold Chile 1892
Zeta 70511	Bk Iron	Glasgow Henderson	1877	639	185 29 17	Swansea Merchant Shipowners Co. Ltd.	

Appendix III

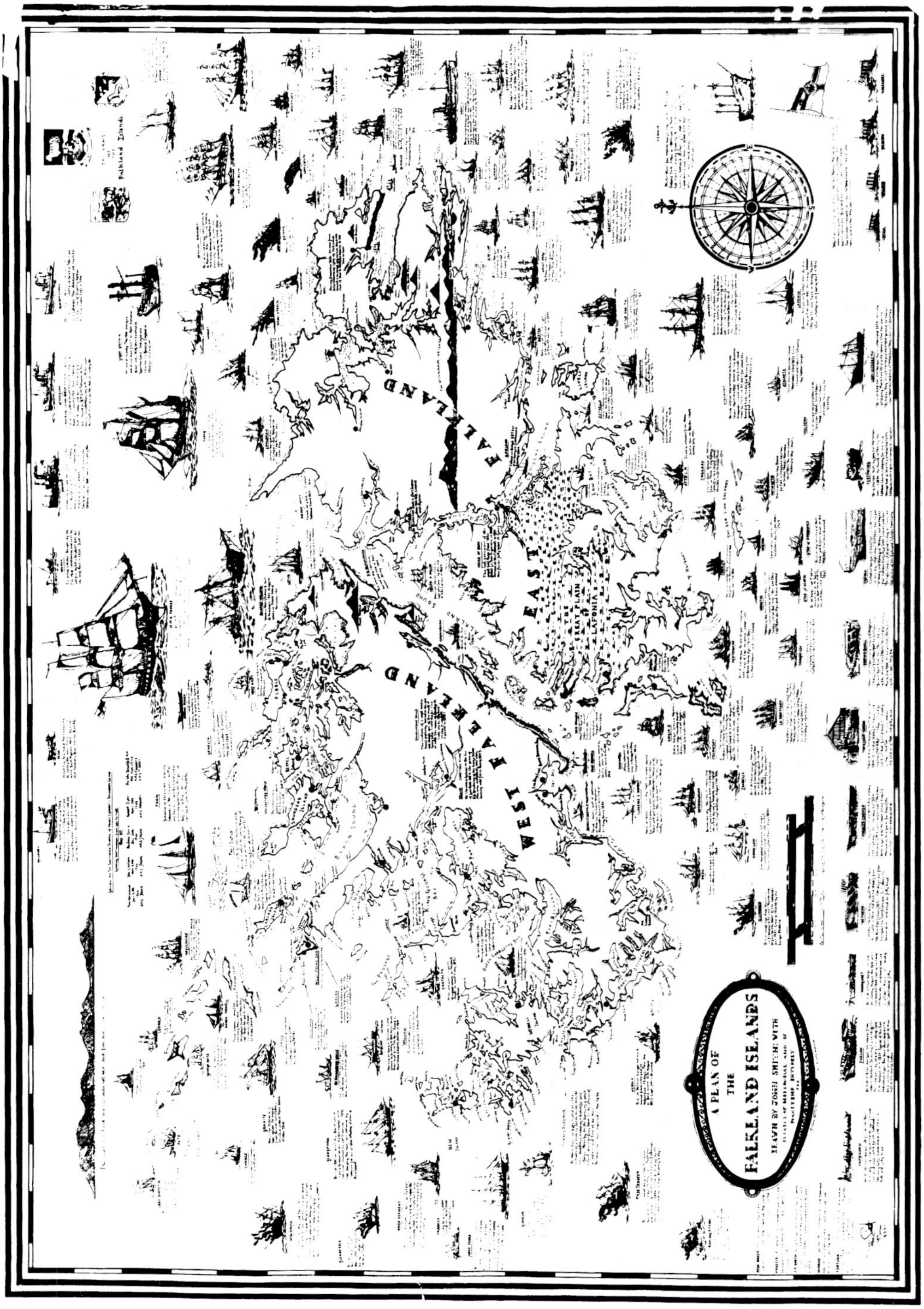

Appendix IV

SANCTIONED BY THE BOARD OF TRADE. MAY 1855.
IN PURSUANCE OF 17 & 18 VICT. c. 104.

SHIPS' REPORTS RECEIVED 6 NOV 71 [illegible] G. R. & R. [illegible]

(O) **OFFICIAL LOG BOOK.** **No. 3.**

FOR

EITHER FOREIGN-GOING OR HOME TRADE SHIP.

Name of Ship.	Official Number.	Port of Registry.	Registered Tonnage.	Name of Master.	No. of his Certificate (if any).
"Delta"	51107	Swansea	537	Peter Rousey	115.38

Date of Commencement of the Voyage 25th September 1871 from Shields to

Nature of the Voyage or Employment Coquimbo Chile Southamerica voyage abandoned in Consequence of vessel getting on shore on stranding [illegible]

Delivered to the Shipping Master of the Port of Swansea on the 31st

day of October 1871.

Signed [illegible]

Shipping Master.

NOTE.—The above Entries are to be filled up by the Master, and the Log Book is to be delivered to the Shipping Master within forty-eight hours after the Ship's arrival, or upon the discharge of the Crew, whichever first happens, in the case of a "Foreign-going Ship;" and within twenty-one days after the 30th of June and the 31st of December respectively in every year in the case of a "Home Trade Ship."—*Sec* 286.

LONDON:
PRINTED BY GEORGE EDWARD EYRE AND WILLIAM SPOTTISWOODE,
PRINTERS TO THE QUEEN'S MOST EXCELLENT MAJESTY.
1869.

The cover of the Official Log Book of the 'Delta' for the voyage commencing 25th September, 1871, to Coquimbo, Chile.

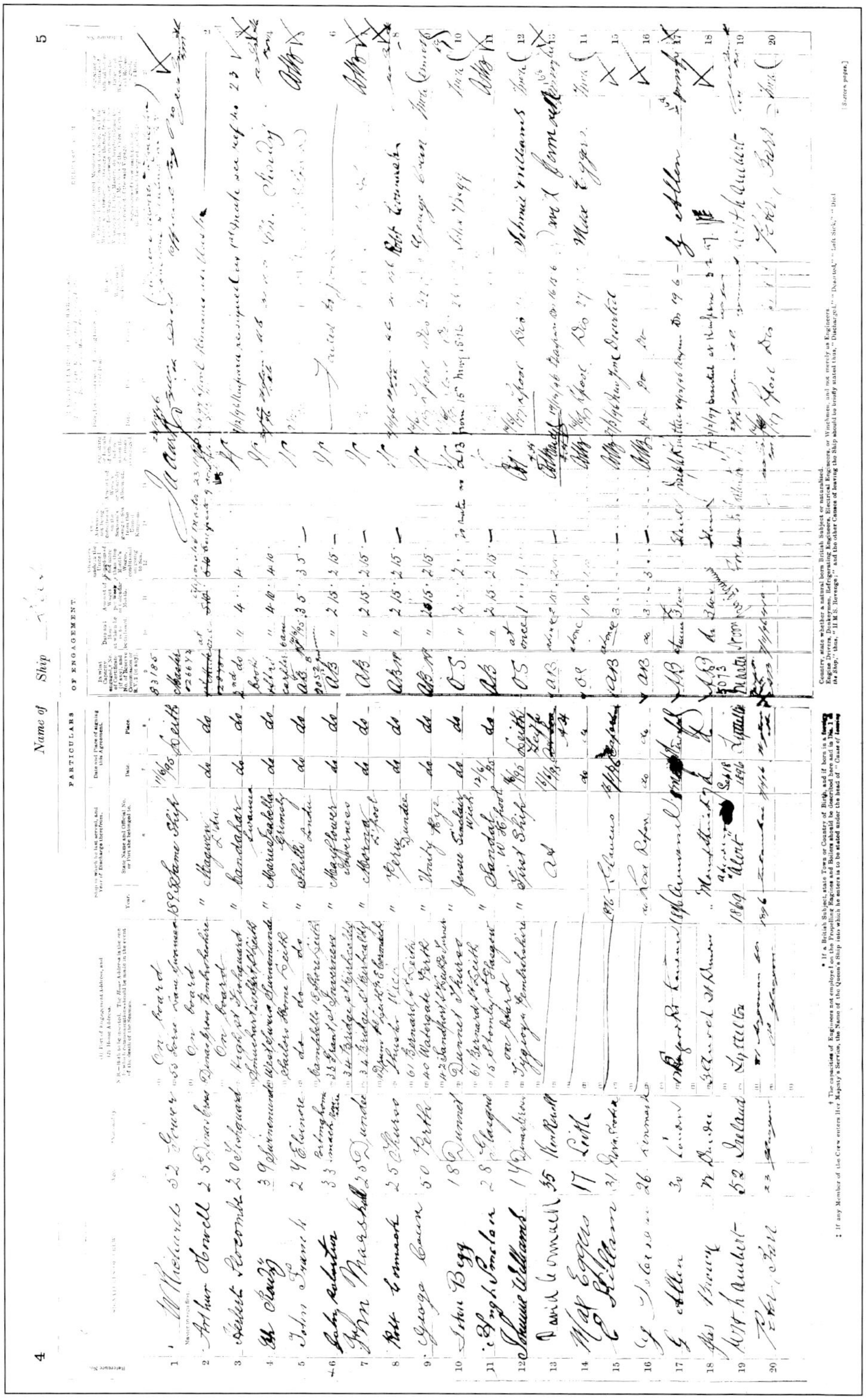

The articles of engagement for the 'Delta' for the voyage commencing 11/6/1895.

List of Publications Consulted

Alban J. R., *Swansea 1184-1984.*

Boorman, D., *The Brighton of Wales.*

Conway Maritime Press (in association with University of Exeter), *The New Maritime History of Devon*, Vol. II.
Clark, A. H., *The Clipper Ship Era.*
Conrad, Joseph – all of his work which relate to the sea.
Chichester, Francis, *Along the Clipperway.*
Campbell-Jones, Susan, *Welsh Sail.*
Chamberlain, Muriel E., *The Grenfells of Kilvey.*
Chamberlain, Muriel E. (edited by), *The Welsh in Canada.*
Colcord, Joanna C., *Songs of American Sailormen.* New York, 1938.
Chadwick, Richard, 'Trading in Ores, 1600-1900' (*Tamar* Journal, 1983).

Davies, J. M., 'The Morris Family' (*Gower* magazine).

Eames, Aled (with Hughes, Emrys):
Porthmadog Ships.
The Twlight of Welsh Sail.
Ships and Seamen of Anglesey.
Ship Master (Pub. Gwynedd County Council Dept. of Culture & Leisure).
Ventures in Sail.
Shrouded Quays (Pub. Gwasg Carreg Gwalch).

Falkus, Hugh, *Master of Cape Horn.*

Greenhill, Basil & Giffard, Ann:
Victorian & Edwardian Ship & Harbours.
Victorian & Edwardian Sailing Ships.
The Merchant Sailing Ship.
Women Under Sail.
Greenhill, Basil:
The Merchant Schooners.
West Countrymen in Prince Edward's Isle.
Greaves, Thomas & Culver, William, *The Americas* (Manchester University Press).
Grant-Francis, Col., *The Smelting of Copper in the Swansea District of South Wales from the Time of Elizabeth to the Present Day (1881).*
Grantham, John, *Iron Shipbuilding* (London, 1858).
Gabb, Gerald, 'The Strand at Swansea in the 1890s' (*Gower* magazine).

Howell, R. G., *Under Sail. Swansea's Cutters, Tallships & Seascapes* (Swansea Museum Services).
Hughes, Henry, *Immortal Sails.*
Hay, K. M. & Roberts, Joy, *The Sea Voyages of Edward Beck in the 1820s.*
Hughes, Emrys & Eames, Aled, *Porthmadoc Ships.*
Hugill, Stan:
Sailortown.
Shanties from the Seven Seas.
Hughes, Gareth, *A Llanelli Chronicle.*
Harris, John, *Essays in Industry and Technology in the Eighteenth Century, England and France.*

Innes, J., *Old Llanelly.*

Jones, W. H., *History of the Port of Swansea.*
De Jong, Nicolas H. & Moore, Marven E., *Shipbuilding on Prince Edward Island* (Canada).
Jones, W. H. S., *The Cape Horn Breed* (New York).

Kooiman, W., *Last of the Guano Traders* (California, U.S.A.).
Kverndal, Roald, *Seamen's Missions; Their Origin and Early Growth* (Pub. by William Carey Library, Pasadena, U.S.A.).

Lubbock, Basil:
The Last of the Windjammers.
The Nitrate Clippers.

Morris, Bernard, *The Houses of Singleton.*
The *Mariners Mirror* – various articles (inc. July 1944 regarding Swansea Pilot Cutters by J. E. Coates).
Morgan, D. W., *Brief Glory.*

National Maritime Museum, *Shipbulding in North Devon.* Monograph and Reports, No. 22, 1976.

Paul, Rodman W., 'The Wheat Trade between California and the United Kingdom' (*Mississippi Valley Historical Review*, Vol. XLV, December 1958).

Ridley, Jasper, *Lord Palmerston.*
Rees, Ronald, *'King Copper' – South Wales and the Copper Trade, 1584-1895.*
Robins, Nigel A., 'Derelict Vessels in the New Cut, Swansea' (*Gower* magazine).
Roberts, R. O., 'The Smelting of Non Ferrous Metals Since 1750' (*Glamorgan County History*) and other works.
Rees, Susan, *In Search of the Bells of Santiago.*

Smith, C. E. G. & Gibson, M. E., 'Yellow Fever in Wales, 1865' (*Medical History*, xxx, 1986.
Sea Breezes – various articles from 1930s onwards.
Shalimar (F. C. Hendry) – all his works.
Slocum, Joshua:
Sailing Alone Round the World.
The Voyage of the 'Liberdade'.
Stevens, *Stevens on Stowage.*
Stammers, Michael K. & Kearon, John, *The Jhelum.*
Stammers, Michael K., *Ship Figureheads.*
Swansea Maritime and Industrial Museum, *Swansea Cape Horners Remembered.*
Swansea Museum Fact Sheets, as below:
1. Chauncey Townsend – Industrial Entrepreneur.
3. John Scott.
4. Notes on Coal-mining in the Llansamlet area.
5. Early Copper Works.
6. Later Copper Works.
7. The Rise and Fall of the Copper Industry.
8. Tinplate.
9. The Grenfells.
10. The Morris Family.
11. The Coppermen.
Canals of the Swansea Area.
Swansea Museum, *Slums*, by Bob Dean, Parts 1-3.
Swansea City Archives:
TC100/Index 15 (Report on the Estates and Property of the Corporation of Swansea, 1850, p.5.
The Bath Family Records.
Swansea newspapers:
Cambrian newspaper – first published 1804; numerous articles. Also in this connection, Mrs Marilyn Jones, Local History Librarian, Swansea, and in charge *Cambrian* index.
The Herald of Wales – various articles.
Swansea Evening Post – various articles.
Western Mail – various articles.
Swansea Museum History Project (edited by Gabb, Gerald), 'Report by George Thomas Clark, Superintending Inspector Swansea General Board of Health, May 1849'.

Thomas, N. L., *The Story of Swansea's Districts and Villages.*
Trott, Clive, 'The Copper Industry' (from *Neath and District.* A symposium published by Elis Jenkins, 1974).

Underhill, Harold, *Deep-water Sail.*

Villiers, Alan:
The Set of the Sails.
The War with Cape Horn.

Webber, Vivian A., *Journal of a Voyage Round Cape Horn, 1859.*
Williams, Sir Glanmor (edited by), *Swansea – An Illustrated History.*
Woolard, Claude L. A., *The Last of the Cape Horners.*

Acknowledgements

IN A WORK OF THIS KIND, one is constantly aware of what others have written before. It is hard, when describing how a pot of jam is made, to find a new way of specifying the ingredients and methods, yet a cookbook might be incomplete without it. I have gone back to original sources wherever possible, but that does not avoid, in some cases, having to say in different words what others have said before me. And importantly, in many instances I am indebted to them for material of which I would not otherwise have been aware. It is in this context that I have separately presented a list of works consulted, and here gratefully record my debt.

In one area, the memorabilia of, by, and relating to, local Cape Horners I was dependant on local response to newspaper and other appeals for assistance, and my own enquiries and approaches to perfect strangers, who often had little time to spare from their busy lives. I am sure that all whom I have met will be aware of my deep sense of gratitude. In particular, I would like to thank those names who appear in the text.

Apart from the above, I would specially like to thank the following:

Mrs Marilyn Jones, Local History Librarian of Swansea and in charge of the 'Cambrian Index', without which this book could not have been produced in its current form.

Mr Michael Lewis, Curator, and Ms Roberta Williams, Exhibition Officer, Swansea Maritime Museum.

Ms Rosalyn Gee, Curator, and Mrs Bernice Cardy, of the Swansea Museum, whose help has been invaluable.

Mr Gerald Gabb, Education Officer, Swansea Museum, whose various books and articles on the local history of Swansea are invaluable to anyone interested in that subject.

Mrs Jennifer Sabine, currently Acting Curator of the Glynn Vivian Art Gallery, Swansea, who has played such a large part in the conception and production of this book.

Ms Susan Beckley, Chief Archivist, Swansea City Archives, and her staff, without whom this book could not have been produced as it now appears before the public.

Mr William G. Thomas, Superintendent, San Francisco Maritime National Historic Park, who kindly gave me of his time.

Irene A. Stachura, Librarian, and Bill Kooiman, Library Technician, of the J. Porter Shaw Maritime Library, California, who so cordially welcomed me and afforded so much valuable advice and material.

Sam Davidson, author of *Samuel Walters – Marine Artist* and *Marine Art and Liverpool Painters, Places, and Flag Codes*, whose authoritative advice on marine paintings and ships' flags has been invaluable.

Mr Dale Thomas of Monmouth whose advice and help has contributed so much, as well as saving me from many pitfalls.

Mr Michael Bevan of Swansea whose enthusiastic research into his family connection with Cape Horners and the history of Gower has been a source of assistance and inspiration to me.

Mr Colin Rees of Mumbles, Swansea, who has provided so much of his family material, and encouragement, and Mrs Susan Rees for permission to quote from her *Bells of Santiago*.

Mr Ronald Rees of New Brunswick, Canada, who kindly let me make use of the research material for his *'King Copper' – South Wales and the Copper Trade, 1584-1895*.

Finally, my deepest thanks to all at Dinefwr Press, Llandybie. This is the second book they have printed for me and words cannot do justice to their helpfulness, editing skill, ability to fulfill targets, and the quality of their work.

Index